CELL DIFFERENTIATION OF NEOPLASTIC CELLS ORIGINATING IN THE ORAL AND CRANIOFACIAL REGIONS

Cell Differentiation of Neoplastic Cells Originating in the Oral and Craniofacial Regions

Toshiyuki Kawakami
and
Hitoshi Nagatsuka

Nova Biomedical Books
New York

Copyright © 2009 by Nova Science Publishers, Inc.

All rights reserved. No part of this book may be reproduced, stored in a retrieval system or transmitted in any form or by any means: electronic, electrostatic, magnetic, tape, mechanical photocopying, recording or otherwise without the written permission of the Publisher.

For permission to use material from this book please contact us:
Telephone 631-231-7269; Fax 631-231-8175   Web Site: http://www.novapublishers.com

NOTICE TO THE READER
The Publisher has taken reasonable care in the preparation of this book, but makes no expressed or implied warranty of any kind and assumes no responsibility for any errors or omissions. No liability is assumed for incidental or consequential damages in connection with or arising out of information contained in this book. The Publisher shall not be liable for any special, consequential, or exemplary damages resulting, in whole or in part, from the readers' use of, or reliance upon, this material.

Independent verification should be sought for any data, advice or recommendations contained in this book. In addition, no responsibility is assumed by the publisher for any injury and/or damage to persons or property arising from any methods, products, instructions, ideas or otherwise contained in this publication.

This publication is designed to provide accurate and authoritative information with regard to the subject matter covered herein. It is sold with the clear understanding that the Publisher is not engaged in rendering legal or any other professional services. If legal or any other expert assistance is required, the services of a competent person should be sought. FROM A DECLARATION OF PARTICIPANTS JOINTLY ADOPTED BY A COMMITTEE OF THE AMERICAN BAR ASSOCIATION AND A COMMITTEE OF PUBLISHERS.

LIBRARY OF CONGRESS CATALOGING-IN-PUBLICATION DATA

Kawakami, Toshiyuki.
  Cell differentiation of neoplastic cells originating in the oral and craniofacial regions / Toshiyuki Kawakami, Hitoshi Nagatsuka.
       p. ; cm.
   Includes index.
   ISBN 978-1-60456-932-2 (softcover)
   1. Mouth--Cancer--Histopathology. 2. Head--Cancer--Histopathology. 3. Odontogenic tumors--Histopathology. I. Nagatsuka, Hitoshi. II. Title.
   [DNLM: 1. Odontogenic Tumors--physiopathology. 2. Cell Differentiation. 3. Cell Transformation, Neoplastic. 4. Morphogenesis--physiology. 5. Receptors, Notch--physiology. WU 280 K22c 2008]   RC280.M6K39 2008
   616.99'431--dc22
                                                                                         2008026978

Published by Nova Science Publishers, Inc. ╪ New York

CONTENTS

Preface		vii
Chapter 1	Introduction	1
Chapter 2	Biology of Neoplasms	3
Chapter 3	Developmental Biology of Teeth	5
Chapter 4	Developmental Oral Craniofacial Biology	9
Chapter 5	Mandibular Bone and Cartilage Development	11
Chapter 6	Notch Signaling in Cell Differentiation and Development	17
Chapter 7	Cell Differentiation in Pathological Conditions	19
Chapter 8	Odontogenic Neoplasms	21
Chapter 9	Notch Signaling in Ameloblastomas and Ameloblastic Carcinomas	23
Chapter 10	Cell Differentiation in Odontogenic Neoplasms	27
Chapter 11	Cell Differentiation in Neoplasms of Bone and Cartilage	33
Chapter 12	Cell Differentiation in BMP-Induced Heterotopic Osteogenesis	37
Chapter 13	Conclusion	41

References 43

Index 51

PREFACE

Development of the oral and craniofacial region is a complex and fascinating set of processes which require a sequential integration of numerous biological steps. For medical and dental doctors, interest is particularly high in this region, because it is composed of three blastoderms - ectoderm, mesoderm, and endoderm - as well as neural crest cells. There are many different types of neoplasms in this region. In general, proliferation, development and cytological differentiation of the neoplastic cells reflect the normal physiological development of the outbreak mother cells and/or tissues. Collected human neoplasm cases, such as osteosarcoma appearing in the oral and craniofacial region, are examined regarding the immunohistochemical expression of some morphogenesis regulation factors. Furthermore, examination of Notch signaling is also conducted for some odontogenic neoplasms. This chapter mainly describes the examination results of some morphogenesis regulation factors, such as Notch signaling, in the neoplastic cells originating in the oral and craniofacial region, especially in the odontogenic neoplasms, in both well-differentiated and poorly-differentiated neoplasms of tooth germ enamel organ-derived neoplasm. In general, these morphogenesis regulation factors are responsible for cytological regulation of cell fate, morphogenesis and/or development. The results suggest that these factors play some role in cytological differentiation or acquisition of tissue specific characteristics in neoplastic cells. Furthermore, there would appear to be a relationship between the cytological differentiation in the oral and craniofacial neoplastic cells and the physiological development and differentiation of their originating mother cells and tissues of the oral and craniofacial region.

Chapter 1

INTRODUCTION

A neoplasm is an abnormal growth mass of cells and/or tissue of which exceeds and is uncoordinated with the physiological cells and/or tissues. Thus, neoplasms are said to be transformed because they continue to replicate, apparently oblivious to the regulatory influences that control physiological original cell and/or tissue growth. In neoplasm, the words "differentiation and anaplasia" are applied to the originating parenchymal cells that constitute the transformed neoplastic components of neoplasms. The word "differentiation" of parenchymal cells refer to the extent to which they resemble their physiological features, both morphological and functional in findings.

Chapter 2

BIOLOGY OF NEOPLASMS

In benign neoplasms, the parenchymal cell components consist of well-differentiated cells that quietly resemble their originating mother cell and/or tissue. In benign neoplasms, mitoses are extremely rare in number and are of normal physiological features. In malignant neoplasms, they are characterized by a wide variety of parenchymal cell differentiation, from completely well- to un-differentiated. Malignant neoplasms are composed of poorly- to un-differentiated cells, when said to be anaplastic. Anaplasia, a lack of differentiation, is thought to be a standard of malignancy. Anaplasia means literally to "form backward". In fact, it is well accepted that a malignant neoplasm originates from stem cells in tissues. Therefore, a failure of differentiation, rather than de-differentiation of physiologically specialized cells, accounts for un-differentiated neoplasms. In general, the more malignant and the more un-differentiated (anaplastic) a neoplasm, the less likely it is to have specialized functional activity. The cells in benign neoplasms are almost always well-differentiated and resemble the physiological normal cells of the originating mother cells and tissue. The cells in malignant neoplasms are more or less differentiated, but some loss of differentiation is always present. The differentiation in these neoplastic lesions is controlled by, in part, the normal physiological regulation system. The regulation mechanism of specialization growth of a physiological organization functions on a neoplasm, and the specialized growth of neoplasms is also regulated. Therefore, it is important to examine the normal physiological cell differentiation mechanisms for understanding the regulation system of neoplastic conditions.

Chapter 3

DEVELOPMENTAL BIOLOGY OF TEETH

Teeth develop physiologically, as deciduous teeth and permanent teeth, from oral ectodermal epithelium and neural crest derived neuroectodermal mesenchymal cells [20]. The enamel is derived from ectoderm of the oral cavity, and all other tissues differentiate from the mesenchyme derived from mesoderm; neural crest cells are imprinted with morphogenetic information before or shortly after they migrate from the neural crest. Tooth development is initiated by the inductive influence of the neural crest mesenchyme on the covering ectoderm. Tooth development is a continuous step. However, it is usually divided into a bud stage, cap stage, and bell stage.

Odontogenesis or tooth development is a complex and highly-regulated process characterized by sequential epithelial-mesenchymal interactions leading to tooth initiation, morphogenesis and cell-differentiation with eventual formation of enamel, dentin and cementum matrices [17,18]. Osteogenesis or bone formation is also a tightly-coordinated process involving many different tissues that interact with each other via a matrix-mediated inductive mechanism, and ending in the formation of a specialized tissue, bone. Both of these processes, though distinct, are closely related in that they share common signaling pathways in terms of morphological differentiation of their cells and functional differentiation of their matrix proteins. The ameloblasts and odontoblats, which are exclusive enamel matrix-producing and dentin matrix-producing cells respectively, share several molecular characteristics with the osteoblast which is the bone matrix-forming cell. Runx2, a transcription factor, is essential for osteoblast differentiation (Figure 1). In the course of odontogensis, the Runx2-knockout mice experiments results suggest as follows: Runx2 is associated with morphogenesis of teeth and matrix protein gene expression [42]. Next, Next, compared to the incisor tooth germ, the molar tooth germ is more strongly subjected to control by Runx2, suggesting the presence of factors involved in odontogenesis of the incisor tooth germ which are different from those present in osteoblasts. Furthermore, in Runx2-knockout mice differences in expression of osteopontin and osteocarcin, matrix proteins common for teeth and bone, suggest different mechanisms of cellular differentiation or transcription regulation pathways in incisor odontoblast and bone forming cells, or osteoblasts [43].

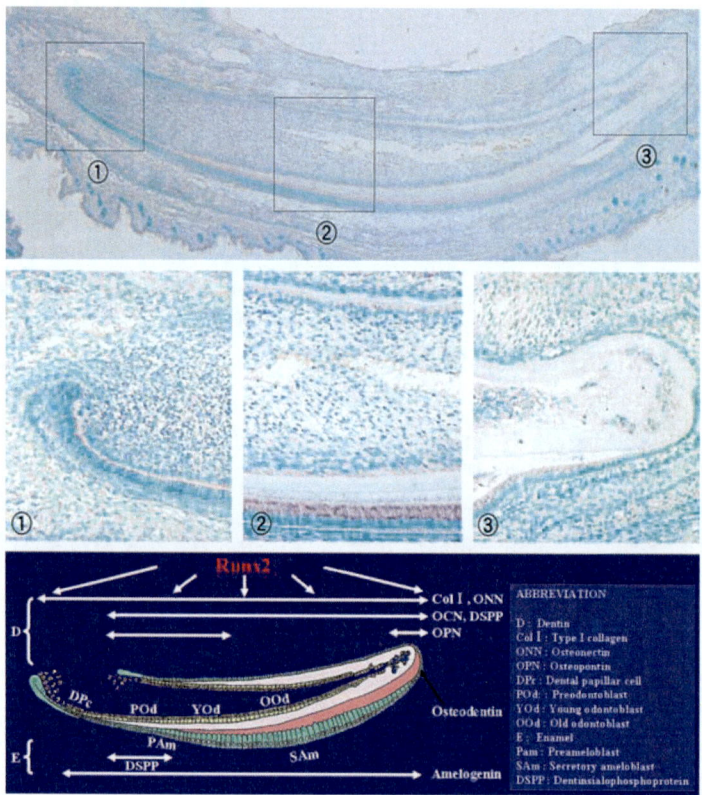

Figure 1. Runx2 regulatory control of the signaling pathway of odontogenesis. The gene expression is visible in various stages of ameloblasts (PAm, SAm, and maturation stage).

The tooth germ basal membrane interposed between the odontogenic epithelium and mesenchyme mediates the sequential and reciprocal epithelial-mesenchymal interactions essential for morphogenesis and cell-differentiation for tooth formation (Figure 2). It composes some isoforms of type IV collagen, laminin, nidogen/entacin, heparan sulfate, proteoglycan, fibronectin, and other components molecules [39,46,80]. The molecules of type IV collagen, a major framework-forming peptide of basal membrane, are heterotrimers composed of three α chains that exist in six genetically distinct forms (α1 to α6) and with at least three molecular forms [48,54,55]. The expression patterns of type IV collagen molecular forms in tooth germ organogenesis and the marked stage-specific changes in the type IV collagen distribution during the odontogenesis are limited [19]. According to the examination results using mouse developing molar tooth germ at the dental placode and bud stage in the course of odontogenesis, the basal membrane of the oral cavity epithelium expresses α1, α2, α5 and α6 chains while the gubernaculums dentis, in addition to above 4 chains, also expresses α4 chain. An asymmetrical distribution of α4, α5 and α6 chains has been observed at the bud stage in the odontogenesis. At the early bell stage, the basal membrane associated with the inner enamel epithelium of molar germ expresses α1, α2 and α4 chains while the basal membrane of the outer enamel epithelium only expresses α1 and α2 chains. With the onset of dentin formation, the collagen α chain profile of the basal membrane of inner enamel epithelium gradually disappeared. From the bell stage, however, the gubernaculums dentis consistently expressed α1, α2, α5 and α6 chains, this distribution pattern resembles the one of the fetal oral cavity epithelium. These features suggest that the odontogenic stage- and the position-specific type IV collagen α subunit distribution is according to the tooth germ odontogenesis, and its changes are essential for the morphogenesis and cell-differentiation for the tooth development [8,39,74].

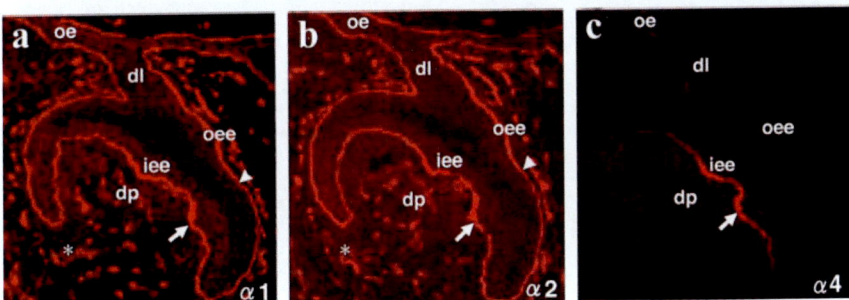

Figure 2. Distribution of α chains of type IV collagen in early bell stage of E15 mice. The basal membrane of inner enamel epithelium (iee, arrow) stains for α1 (a), α2 (b), and α4 (c) chains. oee: outer enamel epithelium, dp: dental pulp, oe: oral epithelium, dl: dental lamina, *: vascular basal membrane

Tenascin is an extra-cellular matrix glycoprotein which appears to regulate cell morphology [40]. It is more restricted to tissue distribution than fibronectin and is able to interface with the cell binding function of fibronecion. Tenascin is most typically expressed in epithelial-mesnchymal interactions in during physiological development and in the stromal tissue of malignant neoplasms. Extra cellular matrix protein is shown to play important roles in cellular growth and differentiation, in complex cell matrix interactions, in physiological organ development and neoplastic transformation course [40].

At first in dental lamina of the bell stage of human tooth germ, tenascin is present only on the submucosal connective tissue side, not on the dental follicle tissue side [6]. At this stage there are no morphological differentiations in the odontogenic epithelium on either side. Concerning fibronectin, a weak or negative localization is seen in the condensed mesenchyme surrounding the dental lamina. In the cap stage, different patterns of the distribution between tenascin and fibronectin is evident in the human tooth germ. Strong tenascin accumulation is present in the dental papilla under the basal membrane, preodontogenic layer and osteogenic tissue of alveolar bone. However, tenascin is immunohistochemically negative in the dental follicle, the fibroblastic layer developing to the periodontium [40]. Intense fibronectin is evident in the alveolar bone. The epithelial components of the tooth germ are imunohistochemically negative for both tenascin and fibronectin.

Chapter 4

DEVELOPMENTAL ORAL CRANIOFACIAL BIOLOGY

Development of the oral and craniofacial region in human is a complex and fascinating set of processes which require a sequential integration of numerous biological progresses. The mechanism has been a source of fascination and an object of intensive examination since early scientific medical studies. Regarding the regulation in the development, biological cellular regulation during the processes contain the entire scale of values ranging from ion to molecular interactions.

The oral and craniofacial regions of a 4-week-old human embryo somewhat resemble these regions of a fish embryo at a comparable stage of development. This explains the former use of the adjective "branchial", which is derived from the Greek word "branchia". The pharyngeal (branchial) apparatus consists of the pharyngeal arch, pouch, groove, and membrane. These embryonic structures contribute greatly to the formation of the oral and craniofacial region. The oral and facial primordial begin to appear early in the 4th week around the big *stomodeum*. Facial development depends on the inductive influence of the prosencephalic and rhombencephalic organizing centers. Five facial primordial, the single frontonasal, the pair of maxillary, and the pair of mandibular prominences appear. The pair of facial prominences is derived from the first pair of pharyngeal arches. The prominences are formed predominantly by the neural crest cell proliferation that migrate from the regions of the neural fold into the arches among the 4th week of the embryonic stage. These cells are the major source of mesenchymal tissue components, which include cartilage, bone, and ligament tissues in the oral and craniofacial regions. Therefore in summary, during the 4th- and 5th-week of the embryonic stage, the primitive pharynx is bounded laterally by pharyngeal arches. The arch consists of a core of mesenchyme covered externally by ectoderm, and internally by endoderm. The original mesenchyme of the arch is derived from mesoderm, and later, neural crest cells migrate into the arches and almost all of their mesenchymal components.

Chapter 5

MANDIBULAR BONE AND CARTILAGE DEVELOPMENT

In the oral and craniofacial region, cartilage characteristics are slightly different from other cartilages of general portions, especially joint cartilage. This includes mandibular chondylar, angular and coronoid cartilages. Mandible is composed of mandibular bone and cartilage [52]. This cartilage is classified as a secondary cartilage together with condyle, coronoid and angle. Formation studies of bone and cartilage in the oral and craniofacial region have been conducted by many researchers [12,57-60]. With regard to mandibular angle and coronoid, however, very few reports have been published [28,61]. Their bone formation patters attracts researchers, suggesting large possibilities for both clinical and histological findings. Mandibular condylar cartilage has bone characteristics which are more significant than its cartilaginous characteristics [62,64].

Regarding the development of the mouse mandibular condylar cartilage, at embryonic day 14 there are no development features, although there is some osteoblastic cell proliferation and a small number of mandibular body bone matrices (Figure 3). At the distal upper portion of the developmental mandibular bone, mesenchymal cell proliferation and condensation with no metacholomasia reaction to toluidine blue (TB) are seen. At embryonic day 15, mandibular condylar cartilage is clearly evident, as a metacholomasia reaction to TB, which is firstly expressed at a middle zone of the proliferating mass. At embryonic day 16, the volume of condylar cartilage grows both in length and width. In this stage, articulation occurs between the mandibular bone and condylar cartilage. At the late embryonic stage, the mandibular condylar cartilage further grows both in length and width, especially at the hypertrophy layer. At the connection area of the mandibular trabecular bone and the hypertrophy layer of the condylar cartilage, endochondral ossification occurs. Furthermore, perichondral ossification occurs at the sheath of condyle. That is, direct bone formation occurs at the sheath of the condylar cartilage. At just before birth, endochondral ossification progresses further and the mandiblar condyle volume grows.

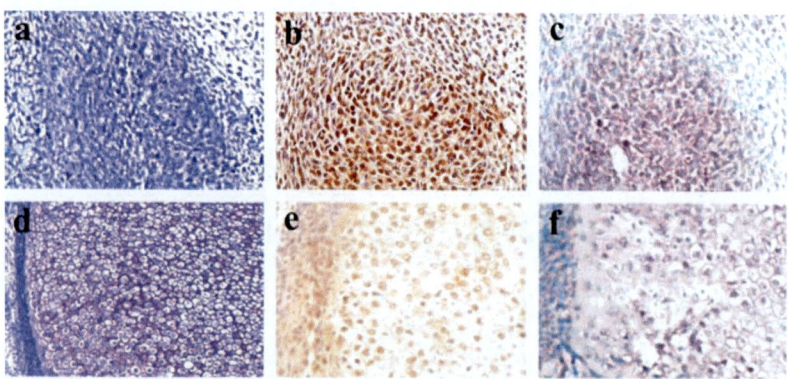

Figure 3. Developmental features of mouse mandibular chondylar cartilage. Mesemchymal cell coagulation of E14 (toluidine blue stain; a) and positive reaction of Runx2 peptide (b) and mRNA (c) are visible. At E18, direct bone formation occurs in the sheath of the cartilage (toluidine blue; d), and Runx2 positive reactions (e) are strongly detected at the sheath with gene expression (f).

The proliferating chondrocytes show immunohistochemically positive reactions to type II collagen, type I collagen and osteopontin. These data suggest that the characteristics of mandibular chondylar cartilage, as secondary cartilage, are slightly different from those of normal physiological articular cartilage. The cartilage takes on the character of bone [37]. In general, Runx2 is a transcription factor necessarily for osteoblast differentiation [21] and bone formation [33].

Mandibular condylar cartilage is recognized as an important growth site and is developed by an endochondral bone formation mode. It is recognized as secondary cartilage, which differs from primary cartilage in morphological and biochemical organization [57]. There are some different components of the extracellular matrix between the primary and secondary cartilage. Immunohistochemical studies for the expression pattern of type I and II collagens [21] have shown that, both types of collagen are simultaneously produced in chondrocytes at this area. Moreover, simultaneous expression of type I and II collagen genes have been confirmed. These findings enable the accumulation of different characteristics and aspect of this cartilage from general endochondral ossification ones. In the reported literature [64], immunohistochemically-positive reactions to osteopontin are detected in almost all layers of the cytoplasm of the mandibular chondylar chondrocytes. The findings from toluidine blue stained specimens of the early developmental stage of mandibular condylar cartilage for this study indicate that, at the distal upper portion of the developmental mandibular bone, mesenchymal cell proliferation and condensation with no metacholomasia reaction are present. Next, mandibular condylar cartilage is clearly evident as a metacholomasia reaction to TB.

Regarding the mandibular angular cartilage, the development starts nearly the same fatal of mandibular chondylar cartilage development. In other words, coagulation of mesenchymal cells have been observed at embryonic day 14, and differentiated to chondrocytes, which show a metachlomasia reaction for TB the next day. After embrionic day 17, endochondral ossification occurs with the invasion of capillaries, and perichondral ossification occurs in the periphery of the cartilage mass. In immunohistochemical examinations, the proliferating chondrocytes of the mandibular angular cartilage show positive reactions to type I collagen and osteopontin, as well as to type II collagen. Therefore, the results show that the characteristics of proliferating mandibular angular cartilage are nearly the same as mandibular condylar cartilage, and differ slightly from normal physiological articular cartilage.

Recently, various studies have shown that mandibular condylar cartilage formation is related to morphogenesis regulation factors and their signaling, such as a fibroblast growth factor receptor, a platelet-derived growth factor receptor [12]. Generally, Notch1 and Math1 are important regulation factors of morphogenesis [64]. There are no reports on mandibular condylar cartilage, although there is a report on the distribution of articular cartilage. Investigations carried out on the two essential factors of expression, Notch1 and Math1 in the mandibular condylar cartilage, reported that, these expression patterns are different from the one in the articular cartilage, and the reactions for Notch1 are present but only localized in the hypertrophic cells [33]. Math1 was distributed mainly in the hypertrophy layer and partially in the proliferate layer. Therefore, the expression patterns of Notch1 and Math1 are slightly different from those of articular cartilage. These results suggest that regulation factors of morphogenesis-Notch1 and Math1-may play some essential role in mandibular condylar cartilage [33]. Since it is inconsonant with the distribution in articular cartilage, it is presumed that the generation of the cartilage and morphogenesis mechanism does not correspond.

In general, Runx2 is a transcription factor necessarily for osteoblast differentiation and bone formation [33]. Furthermore it has been reported that Runx2 regulates chondrocyte hypertrophy during chondrogenesis in long bones. Runx2 is responsible for signaling chondrocyte maturation and endochondral ossification during mandibular condyle advancement. Because matrix protein that characterizes bone, such as type I collagen and osteopontin, has often been expressed, mandibular condylar cartilage has intense bone characteristics. As a result, at embryonic day 14, Runx2 expression is detected by means of immunohistochemical (IHC) and *in situ* hybridization (ISH) examinations, which indicates that the expression leads to the secondary cartilage differentiation. From the data, Runx2 expression is detected at embryonic day 14 as is type II collagen expression. At next embryonic day, type II collagen peptide is expressed. That explains that differentiation has started from embryonic day 14, and that area has turned into cartilaginous tissue the next day. This is in accord with the findings [59], that Runx2 is essential for the onset of formation of the mandibular condylar cartilage, as well as for normal development of Meckler's cartilage, and that muscle tissues influence mandible morphology. This finding supports the above-mentioned research result that Runx2 controls differentiation for mandibular condylar cartilage and induces differentiation [63].

Immunohistochemically, at embryonic day 14, Runx2 peptide is expressed in the nucleus and in the cytoplasm of coagulating mesenchymal cells. Next, the proliferating cells have positive products of Runx2 in cytoplasm and nucleus of almost all coagulating cells. The next day, strongly positive Runx2 reactions are detected in cells of the fibrous and proliferate layers, and weakly labeled in cells of all other layers. Furthermore, Runx2 peptide appears in cells at the sheath of the condylar cartilage. After that, Runx2 factor appear in the cells of the condylar cartilage sheath, and is also distinct in cytoplasm and nucleus. Just before birth, Runx2 positive products are observed in almost all cells of layers, and they are mostly distinct in the sheath of the condyle. Just after birth, Runx2 express in a portion of the hypertrophy cells, especially in their cytoplasm and nucleus.

Proliferating chondrocytes show positive reactions to osteopontin, through the examination periods, particularly in the cytoplasm of the proliferating chondrocytes [62]. At the early stage of the developmental day, weak labels for type II collagen are observed in a portion of coagulating mesemchymal cells. Furthermore, positive products of type II collagen exist in the cytoplasm. Just after onset of development and up to the birth, weak labels for OPN and type II collagen are presented in the deeper layer of the condylar cytoplasm and extracellular matrix [64].

ISH examination of the gene expression revealed that just at onset of development, expressions of Runx2 mRNA appear in cytoplasms of proliferating chondrocytes [63]. After that, Runx2 mRNA is detected throughout almost all cytoplasm of all layers. At just after birth, Runx2 gene expression is observed throughout almost all layers. Furthermore, the signals weakly appear in the upper layers, fibrous and proliferative. The signals are mostly distinct in the cytoplasm. Osteopontin mRNA is detected in the cytoplasm of almost all cells of all layers from onset of development to just before birth. It is distinct in cytoplasm and extracellular matrices. At just after birth, Osteopontin gene signals appear restricted in cytoplasm of maturative and proliferative layers [63].

At the embryonic day of late stage, Runx2 gene expression strongly appears in hypertrophy cartilage, probably due to the differentiation to the osteoblast. This agrees wit a report which explains that the Runx2 expression of IHC and ISH has been identified in the hypertrophy layer, and also takes part in the endochondral ossification mode. Examination results have clearly demonstrated the distribution of Runx2 expression, both of the peptide and its gene, at the cartilage in the sheath of mandibular condyle where direct bone formation is observed. The findings provide evidence for Runx2 control over and/or regulation of perichondral ossification. This is the first time anyone has stated these findings. Because of the differentiation of cartilage, Runx2 expression is generated at embryonic day 14 and the next day, and the displacement of bone from the hypertrophy cartilage induces the expression at just before birth.

In summary, for participation of Runx2 in mandibular condylar cartilage development, there are no development features of mandibular condyle. At the distal upper portion of developmental mandibular bone, mesenchymal cell proliferation and condensation with no metacholomasia reaction to toluidine blue are seen at embryonic day 14. At embryonic day 15, mandibular condylar cartilage is clearly evident as a metachlomasia reaction to TB. Immunohistochemically, at embryonic day 14, expression of Runx2 peptide is observed in the nucleus and the cytoplasm of coagulating mesenchymal cells. After the late stage of the embryonic days, Runx2 factors appear in the cells of the condylar cartilage sheath, and they are also distinct in the cytoplasm and nucleus. In gene expression at embryonic day 14 and 15, expressions of Runx2 mRNA appear in the cytoplasm of proliferating chondrocytes. Days just before birth, the mRNA is detected throughout almost all cytoplasm of all layers. These results suggest that Runx2 plays an essential role for mandibular condylar cartilage development, and that Runx2 is essential for the onset of secondary cartilage differentiation [63].

Chapter 6

NOTCH SIGNALING IN CELL DIFFERENTIATION AND DEVELOPMENT

In general, Notch signaling plays an important role (Figure 4) in the regulation of cell fate, morphogenesis and/or development [5,14]. Regarding tooth development, there are some published data on how the expression of Notch1, 2, and 3 is regulated by epithelial-mesenchymal interactions in the developing mouse tooth and associated with determination of ameloblast cell fate. Jagged1 is also expressed as a ligand of Notch in the developing tooth [35]. Notch signaling is an evolutionarily-conserved cell-to-cell transmembrane interaction mechanism. Furthermore, asymmetric distribution of Notch has been observed in immature cells prior to cell division, suggesting a role in the regulation of daughter cell fate, including whether the cells remain stem cells or give rise to differentiated progeny [36]. Regarding odontogenesis, Notch1 is expressed in stellate reticulum cells, and Jagged1 is expressed in differentiated ameloblasts in the course of tooth development. During tooth development, Notch expression has been associated with the differentiation of odontogenic epithelial and mesenchymal tissues. However, Notch expression is absent in epithelial cells in close contact with mesenchyme, a feature which may be important for ameloblast cell fate. These data suggest that mesenchymal tissue negatively regulates Notch expression in epithelium. In other words, Notch expression is down regulated in odontogenic epithelium juxtaposed to mesenchyme, indicating that odontogenic epithelium needs a mesenchyme-derived signal to maintain the down regulation of Notch [35].

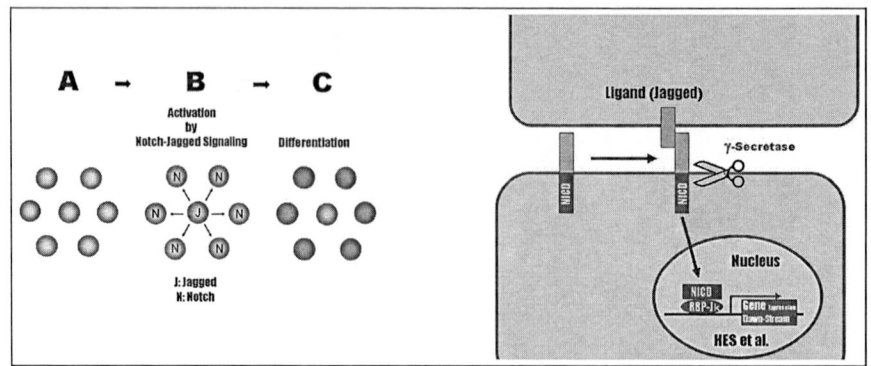

Figure 4. Notch signaling plays in the regulatoion of daughter cell fate.

Chapter 7

CELL DIFFERENTIATION IN PATHOLOGICAL CONDITIONS

In the oral and craniofacial region, there are many types of neoplasms and pathological conditions, such as: odontogenic neoplasms, bone and cartilage neoplasms, pathological bone and cartilage formation and/or proliferation. Especially the odonotogenic neoplasms consist of plural number blastoderms. The neoplastic cell differentiation process is complicated, and it is thought that the cell differentiation and growth pattern are copied from a physiological system of the odontogenesis.

Chapter 8

ODONTOGENIC NEOPLASMS

WHO histological classification of odontogenic tumours

MALIGNANT TUMOURS		Odontogenic epithelium with odontogenic ectomesenchyme,	
Odontogenic carcinomas		with or without hard tissue formation	
Metastasizing (malignant) ameloblastoma[1]	9310/3	Ameloblastic fibroma	9330/0
Ameloblastic carcinoma – primary type	9270/3	Ameloblastic fibrodentinoma	9271/0
Ameloblastic carcinoma – secondary type (dedifferentiated),		Ameloblastic fibro-odontoma	9290/0
intraosseous	9270/3	Odontoma	9280/0
Ameloblastic carcinoma – secondary type (dedifferentiated),		Odontoma, complex type	9282/0
peripheral	9270/3	Odontoma, compound type	9281/0
Primary intraosseous squamous cell carcinoma – solid type	9270/3	Odontoameloblastoma	9311/0
Primary intraosseous squamous cell carcinoma derived from		Calcifying cystic odontogenic tumour	9301/0
keratocystic odontogenic tumour	9270/3	Dentinogenic ghost cell tumour	9302/0
Primary intraosseous squamous cell carcinoma derived from			
odontogenic cysts	9270/3	**Mesenchyme and/or odontogenic ectomesenchyme with or without**	
Clear cell odontogenic carcinoma	9341/3	**odontogenic epithelium**	
Ghost cell odontogenic carcinoma	9302/3	Odontogenic fibroma	9321/0
		Odontogenic myxoma / myxofibroma	9320/0
Odontogenic sarcomas		Cementoblastoma	9273/0
Ameloblastic fibrosarcoma	9330/3		
Ameloblastic fibrodentino–and fibro-odontosarcoma	9290/3	**Bone-related lesions**	
		Ossifying fibroma	9262/0
BENIGN TUMOURS		Fibrous dysplasia	
Odontogenic epithelium with mature, fibrous stroma without		Osseous dysplasias	
odontogenic ectomesenchyme		Central giant cell lesion (granuloma)	
Ameloblastoma, solid / multicystic type	9310/0	Cherubism	
Ameloblastoma, extraosseous / peripheral type	9310/0	Aneurysmal bone cyst	
Ameloblastoma, desmoplastic type	9310/0	Simple bone cyst	
Ameloblastoma, unicystic type	9310/0		
Squamous odontogenic tumour	9312/0	**OTHER TUMOURS**	
Calcifying epithelial odontogenic tumour	9340/0	Melanotic neuroectodermal tumour of infancy	9363/0
Adenomatoid odontogenic tumour	9300/0	see Chapter 1, pp. 70-73	
Keratocystic odontogenic tumour	9270/0		

Figure 5. WHO histopathological classification of odontogenic tumours described in the textbook [11].

Ameloblastoma is classified as a benign, locally-infiltrative odontogenic neoplasm, which is composed of proliferating odontogenic epithelial nests within a fibrous stromal tissue. Some variants have been sub-classified as follows:

solid/multicystic, extraosseous/peripheral, desmoplastic, and unicystic [11]. Furthermore, other variants have been reported in the literature, and these include acanthomatous, ghost cell, and vacuolated or clear cell types [47,51,65-67,76]. Odontogenesis is a complex biological process, and this process is directly reflected in the development of odontogenic neoplasms, especially ameloblastomas [15,16]. It is thought that the above-mentioned variants are due to the developmental complex system [53].

Chapter 9

NOTCH SIGNALING IN AMELOBLASTOMAS AND AMELOBLASTIC CARCINOMAS

Regarding Notch signaling, the focus of our attention is on examining Notch1 and Jagged1 peptide expression, as well as their genes, in ameloblastomas and ameloblastic carcinomas (Figure 6). The speculation on their possible roles is in cytological differentiation and proliferation of ameloblastomas. In one examined case of ameloblastoma, histopathologically, the main specimens showed follicular nests consisting of islands of odontogenic epithelium within a fibrous stroma. Cells of the peripheral layer of these islands were columnar, with hyper chromatic nuclei, and lined up in a palisade fashion whereas the central cells were stellate reticulum-like. Their cytoplasms were generally vacuolated. Some nests showed central cyst formation. In small parts, the odontogenic epithelium exhibited focal basal palisading. Furthermore, occasionally a large number of nests underwent squamous metaplasia with keratinizing pearl formation. In general, degeneration of the parenchymal cells and cyst formation occurred in these ameloblastoma nests. According to the immunohistochemistry examination results, NICD-positive products were detected in most proliferating odontogenic epithelial nests of ameloblastomas by IHC. The positive reactions existed in the cytoplasm and/or nucleus. Strong reactions were seen in the pre-ameloblast-like cells or some localized cells within the nests. In some ameloblastoma nests, there were no positive reactions to NICD. Jagged1 positive reactions were also observed in the cytoplasms of same cell types in the ameloblastoma nests. Strong reactions existed at the peripheral layers. The pattern of distribution and the intensity of expression of Jagged1 were closely similar to the pattern and intensity seen in NICD. Notch1 gene signaling was localized in the cytoplasm of IHC-positive neoplastic cells. These mRNA positive signals showed variable labeling intensity. Jagged1 mRNA signals were also detected in the cytoplasm of ameloblastoma cells, and the strength pattern was nearly the same as that of Notch1. These mRNA signal expressions were not consistent with those of the transcription factor peptides. Histopathologically, the follicular type of ameloblastoma is the most common, consisting of proliferating odontogenic epithelial islands and nests in the fibrous stromal tissues. Cellular modifications, such as squamous metaplasia, keratin pearl formation, parenchyma cell degeneration and cystic changes, may also occur. Morphogenesis is a complex biological process, and this process directly reflects the development and proliferation of neoplasms. Regarding the proliferation of ameloblastomas, some morphogenesis factors are overly-expressed in ameloblastoma tissues in comparison with tooth germs. According to the analysis of gene expression in ameloblastomas and human fetal tooth germs using a cDNA microarray, there are some results are published. That analysis also included tumor-necrosis-factor-receptor-1 (TNFRSF-1), sonic hedgehog (SHH), Cadherins 12 and 13 (CDH 12 and 13), and transforming growth-factor-1 (TGF-β1), the gene expression profile identified candidate genes that might be involved in the origination of ameloblastoma, as well as several genes previously unidentified in relation to human tooth development. The expression of SHH signaling in ameloblastomas, in comparison with human tooth germs is also detected. The literature concluded that the SHH signaling might play a role in epithelial-mesenchymal interactions and cell proliferation in the growth of ameloblastomas [34].

On the morphogenesis factors Notch1 and Jagged1 in ameloblastoma, the results demonstrate that Notch1 (NICD) and Jagged1 are both detected by IHC, and their expression patterns are very similar [81]. This phenomenon means that Notch signaling is activated in the neoplastic epithelium of ameloblastoma. It is likely that the signaling plays the role of daughter cell fate regulation. Positive Notch1 reactions suggest that proliferation and cytological differentiation are probably occurring in these neoplastic cells. This explains the variation in the strength of these signals and their distribution patterns in the ameloblastoma cell nests. Furthermore, the mRNA of Notch1 and Jagged are also expressed in the ameloblatoma cells, as determined by ISH. These mRNA signal expressions are consistent with those of the transcription factor peptides. The examination of larger case series of ameloblastoma and other odontogenic epithelial neoplasms, including epithelial, mesenchymal, benign and malignant entities such as calcifying epithelial odontogenic tumor, adenomatoid odontogenic tumor, keratocystic odontogenic tumor, ameloblastic fibroma, odontoma, and odontogenic carcinoma, would help to elucidate further the role of these genes in odontogenic tumorgenesis. The results suggest that Notch signaling plays a role in cytological differentiation or acquisition of tissue-specific characteristics in these neoplastic cells of ameloblastomas [66].

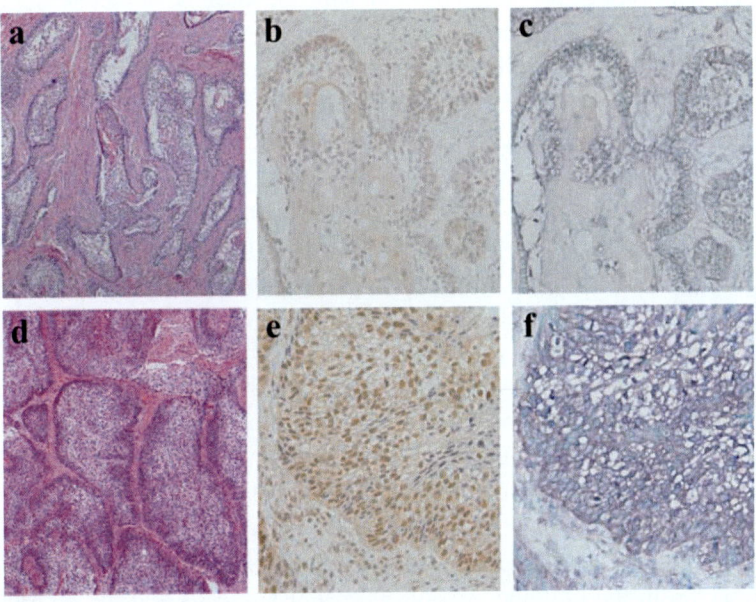

Figure 6. Proliferating follicular nests of ameloblastoma are visible (a). Immunohistochemically Notch peptide is observed in the cells at the peripheral layer (b) and the gene signals are detected in the cytoplasms (c). Polyhedral neoplastic cells are visible in proliferating ameloblastic carcinoma cell nests (d). The Notch positive products are observed uniformly in these cells (e) and the gene expression are also detected in the cytoplasms (f).

Histopathologically, in ameloblastic carcinoma, proliferating polyhedral neoplastic cells show strong cellular atypia, such as mitosis and pleomorphism, especially in peripheral layers of the nests. NICD positive products are observed in most proliferating nests of benign ameloblastoma by IHC, and strong reactions are seen in the cells at the peripheral layer of the nests. In case of ameloblastic carcinomas, positive product are also been detected, and strong reactions uniformly observed. The positive reactions one are comparatively weaker in benign than in malignant tumors. In both benign and malignant cases, the gene (mRNA) expressions have been detected in the cytoplasms of IHC positive cells by ISH. In general, Notch signaling is responsible for cytological regulation of cell fate, morphogenesis and/or development. In examinations conducted by the present authors, IHC and ISH examination results have suggested that Notch signaling plays some role in cytological differentiation or acquisition of tissue specific characteristics in neoplastic cells of tooth enamel organ-derived neoplasms, including benign and malignant neoplasms, ameloblastoma and ameloblastic carcinoma (Figure 6).

Histopathologically, follicular nests proliferate in the fibrous connective tissue in benign neoplasms. In some nests, parenchymal cyst formations or squamous metaplasia are evident. The histopathological features of the malignant neoplasms are as follows: Proliferating polyhedral neoplastic cells show strong cellular atypia, such as mitosis and pleomorphism, especially in peripheral layers of the nests. Notch intra-cellar domain positive products are observed in most proliferating nests of benign neoplasms by immunohistochemistry. Strong reactions are seen in the cells at the peripheral layer of the nests. In malignant ones, positive products are also detected, and strong reactions uniformly observed. The positive reactions in benign neoplasms are comparatively weaker than in malignant ones. In both the benign and malignant cases, the gene expressions are detected in the cytoplasm of immunohistochemistry positive cells by ISH. Our examination results suggest that Notch signaling plays some role in cytological differentiation or acquisition of tissue specific characteristics in neoplastic cells. Furthermore, there would appear to be a relationship between the cytological differentiation in the oral and craniofacial neoplastic cells and the physiological development and differentiation of their originating mother cells and tissues of the oral and craniofacial region.

Chapter 10

CELL DIFFERENTIATION IN ODONTOGENIC NEOPLASMS

Amelogenin is a typical enamel matrix protein. The expression pattern of amelogenin genes (AMGX, AMGY) has not yet been identified in ameloblastomas. In surgical materials, amelogenin gene is expressed in all ameloblastoma cells. The mRNA of AMGY expression increases, although that of AMGX is not. This is an interesting feature in physiological normal male tooth development, in which the expression of AMGY is considerably lower than that of AMGX. This finding suggests that epigenetic change of sex chromosomes may have some correlation with tumorigenesis of ameloblastoma [77].

Regarding the collagen subunits of basal membrane components of oral neoplasms, there are some published data [2,69-71]. Regarding ameloblastomas, co expression of type IV collagen α1 and α2 chains appear as thin lines with limited areas of discontinuity along the basal membrane of neoplastic cell nests [44,45]. The expression staining is strong and in a linear continuous manner, in the periphery of the nests of the desmoplastic types. In the neoplasms, α5 and α6 chains are co-localized as continuous linear patterns demonstrating the tumoral nests from the surrounding connective tissue stroma. These collagen subunits also appear as random intracellular staining of the neoplastic cell nests [9,10]. There is no remarkable differentiation of the distribution pattern among the tumor growth patterns and various cellular subtypes within ameloblastomas (Figure 7). Distribution of α subunit of collagen in the basal membrane of ameloblastic fibromas is uniformly demonstrated in its pattern. Subunits of α1/a2, α4 and α5/α6 are distributed as liner continuous patterns that compartmentalize the neoplastic epithelial cell nests, islands and strands in manner, from the surrounding dental papilla-like ectomesenchymal cell proliferation. These α subunits of collagen are randomly expressed in the periphery preameloblast-like and central stellate reticulum-like cells.

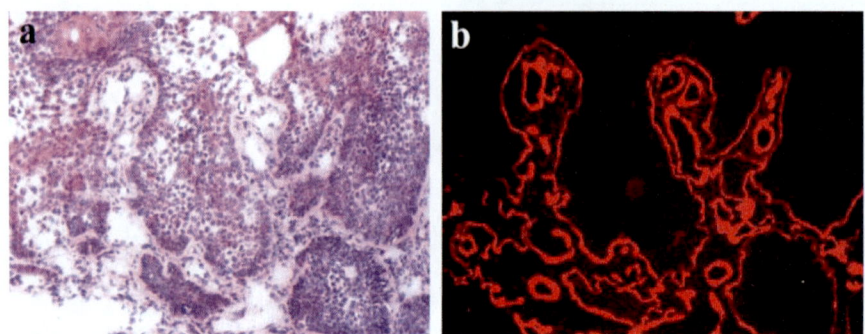

Figure 7. Histopathological feature of examined plexiform ameloblastoma (a) and immunofluoresence localization of type IV collagen α1 chain in the basal membrane zone (b).

In adenomatoid odontogenic tumors, subunits of α1/α2 and α5/α6 are strongly expressed at the area of interface between tumor cells and stromal tissues, especially in the cribriform regions. Faintly to non-positive expression of these collagen molecules is detected in the basal regions of conglomerated masses of solid epithelial whorls/rosettes/nests and duct-like structures. There are intensely positive reactions to the amorphous deposits; however, there is little or no reaction to the mineralized bodies. In malignant neoplasms, at first in ameloblastic fibro-odontosarcomas, α1/α2 and α4 chains demonstrate moderate intensity along with the periphery of the epithelial components, while α5/α6 chains are strongly co-distributed as continuous linear patterns demarcating the benign neoplastic cell nests from the surrounding sarcoma tissues. In the inductive dental hard tissue regions, no reactivity is found. In the malignant neoplasm type of ameloblastic carcinomas, collagen IV α chains demonstrate an irregular and disrupted expression pattern with specific loss of α1/α2 chains. In those regions containing poorly differentiated neoplasm cell nests, there is complete disappearance of α chain subunits. In α5/α6 subunits, there is a discontinuous and fragmented pattern. In primary intraosseous carcinomas, the expression pattern is similar that of ameloblastic carcinomas. From the above mentioned findings for cell differentiation in various benign and malignant odontogenic neoplasms, the basal membrane pattern of neoplastic epithelial cell nests yields three features: (1) The basal membrane of benign and malignant odontogenic neoplasms has distinct α chain subunits of collagen type IV; (2) modifications in the relative abundance of collagen type IV α chains in basal membrane of odontogenic neoplasms probably represent a host protective response, (3) early specific loss of α1/α2 chains proceeds the loss affecting α5/α6 chains during odontogenic neoplasm progression. Therefore, these results suggest that modification and remodeling of basal membrane collagen type IV α chains are dynamic processes crucial for odontogenic neoplastic cell growth and progression [7,45].

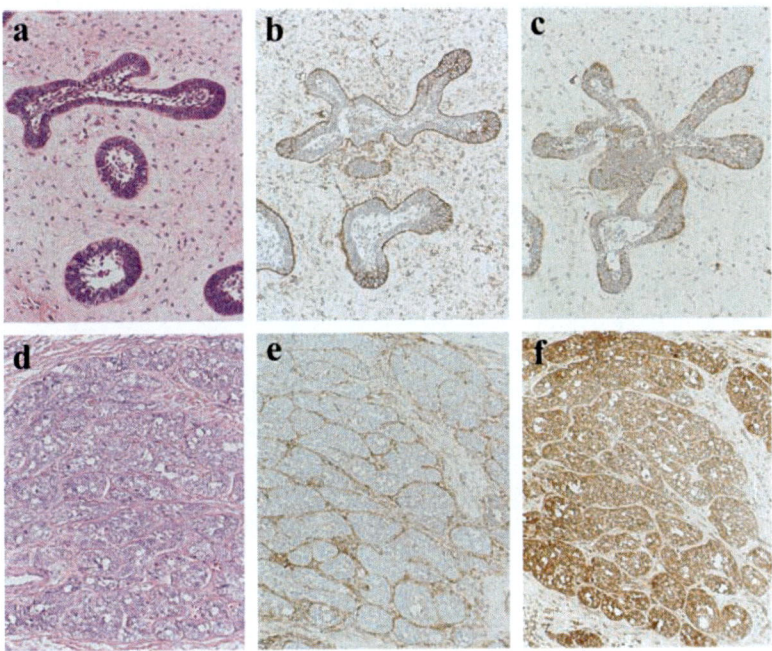

Figure 8. Ameloblastic epithelial islands within scattered dental papilla-like ectomesenchymal tissue (a). HS and heparanase immunohistochemical reactivity accentuates the cellular processes of dental papilla-like cells (b and c). Note the clear confinement and continuity of basal membrane despite the prominent localization of heparanase (b). Histopathologically cancer cell nests of ameloblastic carcinoma with increased nuclear-cytoplasmic ratio and prominent nuclei (d). HS is not detected in the cancer cells but is strongly localized in the stromal tissues adjacent to cancer nests (e). Intense and diffuse heparanase expression is observed in all cancer cells (f) in contrast to the strong staining limited in the basal cells of benign ameloblastoma (c).

Heparan sulphate (HS) and heparanase appearing in the odontogenic neoplasms are interesting molecules for these neoplastic transdifferentiation

[13,50]. HS proteoglycans (HSPG) constitute a group of ubiquitous extracellular matrix macromolecules and are composed of a core protein and covalently linked HS sugar chains (Figure 8).

Although, HSPG plays critical functions in cell-to-cell and cell-to-matrix interactions through core proteins, their HS chains confer most of biological functions [3]. The negatively charged HS chains can bind and sequester numerous heparin/HS binding molecules, including growth factors, cytokines and cell adhesion molecules. HS chains also take part in the important cellar events conferred by these tethered molecules and have an influence on various developmental and pathological processes, such as neoplastic transformation, its local invasiveness and transdifferentiation [3,4]

In the examination results on ameloblastomas, HS is clearly evident on the cell surface of peripheral basal cells and also in the intercellular region of some parabasal cells, while it is not present in the central stellate cells of ameloblastoma nests. Heparanase is expressed in peripheral epithelial cells of ameloblastoma nests. The strong expression is localized at the budding region of the strands mainly in the invasive fronts in histopathological specimens. In adenomatoid odontogenic tumors, the strongly localized and limited expression is present on the surfaces of dark cuboidal cells surrounding the whorls and solid tumor cell nests. HS is also evident in luminar surfaces of some duct-like structures. In the lumen and/or duct-like structures, eosinophilic materials are HS positive. The periphery of immature calcified materials is positive to both HS and heparanase; however, completely calcified materials are negative.

In ameloblastic fibromas, as one of typical epithelial and mesenchymal mixed odontogenic neoplasms, HS exists in nearly the same location as in ameloblastomas. HS is evident in the basal membrane, but is more defined. HS is also present in the ectomesenchymal cells, unlike the stromal cells in ameloblastomas. HS and heparanase are diffusely present both in epithelial and mesenchymal tissue of the neoplasms. In some parts, heparanase exists focally in nuclei of mesenchymal cells. In ameloblastic carcinomas as malignant odontogenic neoplasms, neoplastic cells are absent to HS in contrary of benign neoplasms, such as ameloblastomas. Instead, HS clearly and strongly is present in the stromal tissues, especially in the intercellular matrices within the vicinity of neoplastic cell nests. Regarding heparanase activity, positive reactions are intense and diffuse, and occur in intracellular spaces. The above-mentioned findings are compared with the data of the physiological tooth development, in both experimental animals and human materials. Furthermore, when the results are examined using various types of oral squamous cell carcinoma, there are some differences between the types of varies histological and clinical malignancy grades. In summary, the general localization of HS and the heparanase activity in odontogenic neoplasms are temporally regulated in relation to cellular growth and function. Furthermore, heparanase over-expression is reported to promote hair follicle morphogenesis and its growth [82]. Both hair follicular morphogenesis and odontogenesis are governed by similar growth factors and signaling pathways. Heparanase may also have physiological function in tooth development through local modulation and release of HS-bound growth factors. Taken together, the facts suggest that heparanase may have physiological function in tooth development, and the increase in heparanase expression maybe an important initiating factor for odontogenic neoplastic transformation. The stromal HS sugar molecule localization and heparanase over-expression may represent the malignant progression of ameloblastoma to ameloblastic carcinoma [41].

In odontogenic neoplasms, both benign and malignant, the immunohistochemical distribution of tenascin and fibronectin is compared with that in human tooth germs [72,73]. In ameloblastomas, the extracellular matrix components of the stromal tissue of ameloblastomas exhibit considerable variety: dense and loose connective tissues, hyalinization regions, and stromal cystic spaces. In the hyalinised stroma, tenascin and fibronectin exhibit both positive and negative reactions. In cystic spaces, positive reactions of tenascin and fibronectin are seen. In follicular type of ameloblastomas, the basal membrane region reacts irregularly positive to tenascin. The fibronectin reactivity exhibits uniformly and weakly positive in the dense connective tissue of the stromal region in the follicular ameloblastomas. A partial accumulation of tenascin is found in the basal membrane. The tenascin-positive basal membrane shows fuzzy fibrillar materials, whereas the loose or myxomatous tissues of stromal region of follicular type of ameloblatoma exhibit no reaction to fibronectin. Regarding malignant odontogenic neoplasms, the data on ameloblastic carcinomas is as follows: The stromal tissue and basal membrane of ameloblastic carcinomas show an irregular and strong immunohistochemical-positive reaction to tenascin. In the epithelial cell islands of ameloblastic carcinomas, a scattered or granular positive reaction is evident. The connective stromal tissue of ameloblatic carcinomas shows an irregular and strong reaction to fibronectin. According to the localization pattern of tenascin and fibronectin in the varied types of odontogenic neoplasms, such as benign and malignant, the dental follicle of the tooth germ lacks tenascin but has fibronectin. The osteogenic tissues generally contain both tenascin and fibronectin. The ameloblastic fibromas show positive or negative distributions in the stromal tissues, which suggest a differentiation to the papilla of the tooth germ. It is also suggests that the stromal tissue cells of ameloblastic fibroma differentiate to dental follicular tissues. The relative distribution of menisci and fibronectin can be a marker in histological diagnosis of periodontal and osteogenic fibrous tissues. Furthermore, the findings also suggest that fibronectin and tenascin may be used as markers in cell-differentiation of epithelial-mesenchymal interactions during tooth development and in odontogenic neoplasm for trans-differentiation [40].

Cytokeratins with intermediate filaments characteristic of epithelial cells are very stable and range in weight from 40 to 67kD [42]. Regarding the distribution patterns of cytokeratins in some types of ameloblastomas, such as follicular ameloblastomas, the peripheral columnar cells resembling preameloblasts react positively for NSE-K (52.5kD) and 19-K (40kD) in a linear pattern along the basal membrane, but not for the markers of squamous cells, SE-K (56, 56.5, 58 and 68kD). The reaction pattern of NSE-K and 19K shows a frame-like structure in cytoplasm in peripheral columnar cells, while the central stellate reticulum-like cells show an immunohistochemical reaction with all markers of the cytketain SE-K, NSE-K and 19-K throughout the cytoplasm. In plexiform ameloblastomas, both the central spindle and peripheral cuboidal cells demonstrate a positive reaction with SE-K and 19-K, but not with NSE-K. Regarding the oral mucosa and on the developing tooth germ, there are differences for immunohistochemical reactivities of cytokeratins between the fetal oral mucosa and adult gingiva. The cells of both the fetal oral mucosa and adult gingiva demonstrate positive reactions to SE-K, and negative reactions to NSE-K. However, positive reactivity to 19-K is noted only in cells of the fetal mucosa and Merkel cells in adult gingiva. The dental lamina connecing with the basal cells of the oral mucosa have all these cytokeratins, especially 19-K. In the enamel organ, however, the immunoreactivity for SE-K, a maker of squamous differentiation, is different in each cell layer: positive in the outer enamel epithelium; slight positive to negative in the stellate reticulum and stratum intermedium; and negative in the columnar inner enamel epithelium. NSE-K and 19-K are evident in all types of cells that compose the enamel organ proper. The inner columnar enamel epithelium especially expresses a diffuse positive reaction for NSE-K and 19-K throughout the cytoplasms. These features suggest that the characteristics of the cells of the plexiform ameloblastoma are similar to the fetal oral epithelium not odontogenic epithelium. This later feature suggests that the different expression patterns of cytokeratin in ameloblastoma depend on the follicular or plexiform types. As a coincidence of cytokeratin and functional pattern is not noted among the ameloblastomas and tooth germs, these data suggest that the columnar cells of the follicular ameloblastoma have little resemblance to ameloblast-like cells in cytokeratin structure or in cellular functions [42].

Chapter 11

CELL DIFFERENTIATION IN NEOPLASMS OF BONE AND CARTILAGE

Osteogenesis is a complex biological process, which includes recruitment of stem cells, proliferation of progenitor cells, differentiation of osteoblasts and production and assembly of bone matrix. The different steps of this process are controlled and regulated by multiple local and systemic factors, such as morphogenesis regulators. Therefore, knowledge of the complex interaction between these factors and their contribution to the development of neoplastic osteogenesis is necessary for understanding the characteristics of the neoplasms [29,30].

Histopathologically, in serious cases of osteosarcomas, spindle-shaped sarcomatous cells proliferating mesenchymal tissue directly produce neoplastic osteoid and/or coarse immature bone tissues, while variable histopathological patterns are seen in specimens of some cases. There are mainly osteoblastic and osteoid and/or immature bone matrices, as well as some spindle-shaped fibroblastic neoplastic cells. Osteoblastic neoplastic cells, located around the numerous small osteoid tissues, are comparatively monotonous, varying in size and in shape, and showing hyperchromatic nuclei and mitosis [27].

Immunohistochemically, osteopontin (OPN) peptide, as control, is expressed in almost all cells of the examined osteosarcoma. The strong expression area of OPN is in the comparatively well-differentiated regions of the osteosarcoma, the osteoblastic area containing osteoid tissues. In contrast, weak reaction products are detected in the monotonous spindle-shaped cell proliferation area. Runx2 peptide expression appears in the cytoplasm of almost all neoplastic cells (Figure

9). The expression pattern showed uniformly in the proliferating cells of almost all cases. At the bone and/or well-differentiated osteoid forming region, the positive reactions of Runx2 are slightly strong compared with other poorly-differentiated regions. Regarding the expression of NICD peptide, the peptide is detected in the cytoplasm of neoplastic cells of the comparatively well-differentiated areas of osteosarcomas, which are osteoblastic and chondroblastic containing osteoid and/or chondroid tissues, and this area is the same as the immunohistochemically strongly-stained area by OPN [49]. No expression of NICD peptide is detected in the fibroblastic and poorly-differentiated area. Delta peptide appearance is nearly the same that of NICD peptide. The positive products of Delta appear in the cytoplasms of osteoblastic and chondroblastic cells, but there are no positive reactions in the poorly-differentiated fibroblastic cell proliferation regions. On the other hand, there is no positive reaction immunohistochemically detected in negative control slides.

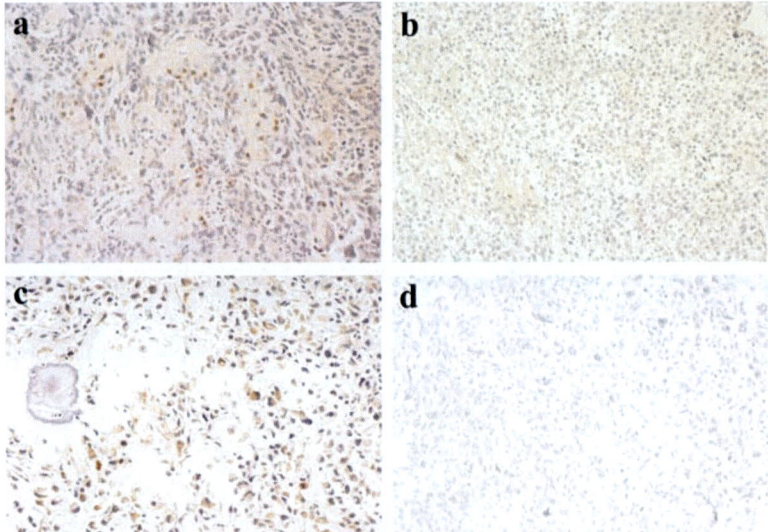

Figure 9. Runx2 positive reaction is visible in the well-differentiated area of osteosarcoma tissue (a) and weakly positive products are detected in the poorly-differentiated area of osteosarcoma tissue (b). NICD is strongly detected in the well-differentiated area of osteosarcoma tissue (c) and no-immunohistochemical reaction of NICD in the poorly-differentiated area of osteosarcoma tissue (d).

In general, it is important to examine the expression or localization of morphogenesis regulators to the neoplastic proliferating conditions, such as benign and malignant tumors. The expression of NICD in an Indonesian male case of osteosarcoma of the maxilla [30] has been reported. As mentioned above, the expression situation of regulation factors of morphogenesis is carried out. Regarding the relationship between these regulation factors and bone tissue, there are some reports in the literature. First, NICD is one of the important regulation factors of morphogenesis. NICD has been reported as a unique and interesting regulator for treatment of osteoporosis. Furthermore, some papers have considered NICD and bone tissue, especially the differentiation of bone forming cells [56,78]. Functional involvement of NICD in osteoblastic cell differentiation has been also reported. However, it is unclear whether Notch1 ligand Delta also induces an identical cellular response in these differentiations. Critical regulation of osteoblastic cell differentiation by Delta-activated Notch1 signaling has also been reported.

The molecular basis for inverse relationship between differentiation and oncogenesis is unknown. However, regarding Runx2, a master regulator of osteoblast differentiation belonging to that runt family of tumor suppressor genes, is consistently disrupted in osteosarcomas. Reports in the literature [75] have described that physiological coupling of osteoblast differentiation to cell cycle withdrawal is mediated through Runx2, and the process are disrupted in osteosarcoma. Furthermore, it has been reported that Runx2 is expressed constitutively in all pathology specimens of human osteosarcoma [1], and expression of Runx2 appeared in the cytoplasm of almost all neoplastic cells of examined cases, and the expression pattern showed uniformly in the proliferating cells of almost all cases. At the bone and/or osteoid forming region, as well at the differentiated area, the positive reactions of Runx2 are slightly strong in comparison with other poorly-differentiated regions. These immunohistochemical results are consistent with the above mentioned discussion. In the present investigation, the NICD peptide is expressed in the area of comparatively well-differentiated areas of osteosarcoma, as well as osteoblastic and chondroblastic areas containing osteoid and/or chondroid tissues. The results are also similar to those of a previously published Indonesian case [30]. With OPN as control peptide in this examination, expression is also detected in almost all cells; the strength pattern of OPN expression is similar to that of NICD. Therefore, Notch peptide maybe closely related to cytological differentiation or acquisition of tissue specific characteristics in neoplastic cells in osteosarcomas.

In summary, the expression of Runx2, NICD, Delta and OPN are examined in neoplastic cells in cases of osteosarcoma. The immunohistocemical expression of Runx2 appears in the cytoplasm of almost all neoplastic cells of examined cases. However, NICD appears in the localized comparatively well-differentiated areas. No expression of NICD peptide is detected in the poorly differentiated area. Delta is shown nearly the same as NICD. Expression of OPN as control appears in almost all cells and the strength of expression is shown in the area of comparatively well-differentiated tissues. Therefore, these results suggest that Runx2, Notch1, and Delta are closely related to cytological differentiation or acquisition of tissue specific characteristics in these neoplastic cells of osteosarcomas.

Chapter 12

CELL DIFFERENTIATION IN BMP-INDUCED HETEROTOPIC OSTEOGENESIS

In general, it has been stated that BMPs, when implanted in heterotopic sites, induces undifferentiated mesenchymal cells to become chondrocytes in the first stage [31,32]. These cells are replaced by bone in a manner similar to that of physiological endochondral (indirect) ossification mode. However, it has been suggested that BMP-induced bone occurs through endochondral-like ossification patterns which differ from those in the physiological normal endochondral ossification process. On the other hand, intramembranous (direct) ossification has also been observed in some cases. There are two types of ossification modes in BMP-induced heterotopic osteogenesis: intramembranous (direct) and endochondral (indirect). However, the nature of osteogenesis has not been clearly detailed [22,23].

Although there is no direct evidence, we believe that the cells inviolved in "chondroid bone", experimentally induced by BMP in mice, temporarily express cartilage phenotypes. They then change directly into bone-forming cells which survive in the "chondroid bone" until the tissue is resorbed and remodeled by true bone tissue. Regarding the expression of TGF-β, in the physiological endochondroid ossification mode, the TGF-β peptide appears in the final-differentiated hypertrophic chondrocytes, and at this stage resorption and replacement by bone occur. We found that the peptide was expressed only in some chondrocites of the earl phase of BMP-induced heterotopic "transchondroid bone formation" in mice. Furthermore, the mRNA was expressed in the same chondrocytes in the early phase of the osteogenesis [23].

Therefore, it is believed that a third ossification mode "transchondroid bone formation" is displayed in BMP-induced heterotopic osteogenesis [24,25]. Chondroid bone, a tissue that has characteristics of bone and cartilage, is formed mainly in BMP-induced heterotopic osteogenesis.

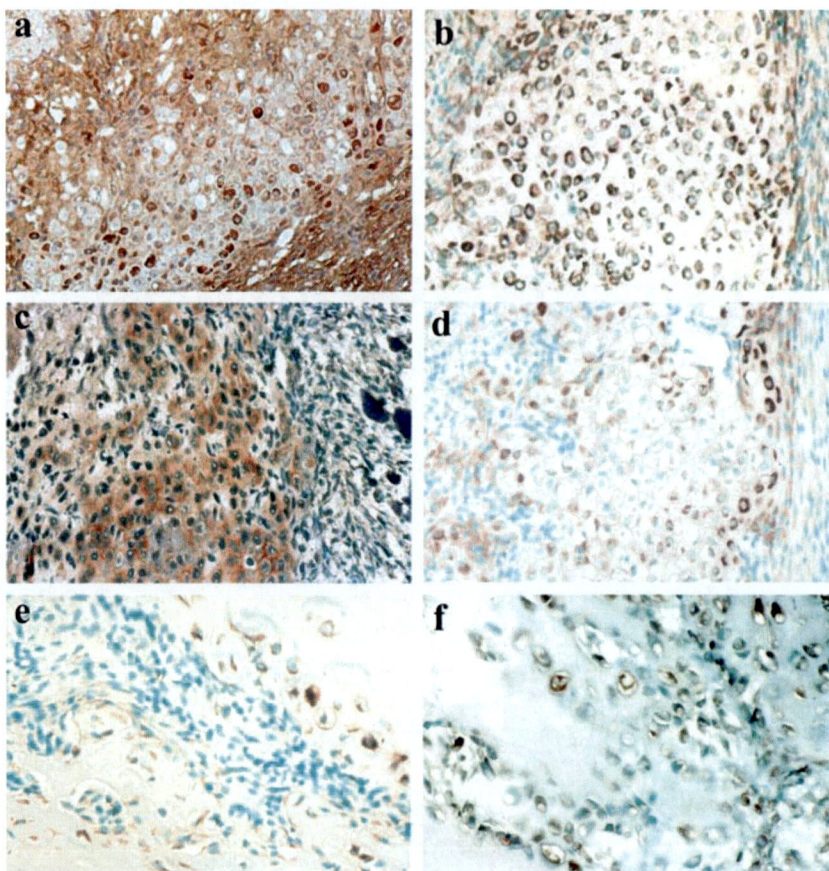

Figure 10. Immunohistochemical localization of type I collagen is observed in the 7-day proliferating cells (a) and the mRNA expression is visible in the cytoplasms (b). Positive immunostainings of type II collagen are distributed in the cytoplasms and matrices (c) and the gene expression is detected the cytoplasms (d). Osteocalcin mRNA signals are visible in the cytoplasms of proliferating cells (e). TGF-β mRNA expression is apparent in some proliferating cells (f).

The experimental results of a BMP-induced model using ddY mice are as follows (Figure 10): Histopathologically in 3-day specimens, spindle-shaped mesenchymal cells proliferated site, and the matrices are stained slightly by HE in 5- and 7-day specimens. Evidence of the proliferation of undifferentiated cells, having cytoplasm and resembling poorly-differentiated chondrocytes, can be seen at the periphery. Within 10 days, perichondral ossification has occurred, and the peripheral matrix of the cluster of cartilage has changed to chondroid bone, connected to perichondral ossification sites. Histochemically, in Mallory's azan-stained specimens, perichondral ossification sites are stained deep blue, although areas stained pale blue are visible in chondral tissue in 10-day specimens. In 10-day specimens stained with toluidine blue, ortho-metachromasia reaction occurs in the matrix, but the matrix of perichondral ossification sites does not occur [26]. In 2- and 3-week specimens, some chondroid tissues displaying ortho-metachromasia are evident in trabecular bone. Immunohistochemically, in 5-day specimens, chondro-osseous formation occurs and type-II collagen is positive in and around chondrocyte-like cells in the chondro-osseous matrix. In 7-day specimens, type 1I collagen is positive in the proliferating spindle cells. Type I collagen is a typical bone matrix protein, and the products of the staining reaction are seen in chondral cells. Type II collagen is also evident in the same cells. These specimens are immunostained for osteocalcin (OCN). Positive staining is observed in 7-day specimens, in particular, and OCN-positive proliferating cartilage cells are clearly visible. In 2- and 3-week specimens of lamellar bone with bone marrow, collagen type II protein disappears from the fibrillar bone matirix. OCN-positive reactions are detected in chondroid cells in 2-week specimens. Regarding these gene expressions in 3-day specimens, type II collagen mRNA are noted in the implanted sites, while OCN mRNA and type II collagen mRNA are undetectable. In 5-day specimens, both type II collagen mRNA and OCN mRNA are detected in some chondrocyte-like cells. Furthermore, in 7-day specimens, type I and II collagen mRNA and OCN mRNA are located in osteoblasts and young osteocytes in osteoid tissues. OCN mRNA is also detected in chondrocyte-like cells, osteoblasts and young osteocytes in osteoid tissues. In 2- and 3-week specimens of lamellar bone with bone marrow formed in the implanted sites, both OCN mRNA and type II collagen mRNA are detected within the newly formed bone.

Recently, a third ossification mode has been proposed: "transchondroid bone formation". Chondroid bone is formed directly by chondrocyte-like cells in the ossification site [24,25]. This form of bone has attracted limited attention, since it was the first reported several years ago. Some hypertrophic chondrocytes are believed to undergo further differentiation into osteoblast-like cells, participating in initial bone formation [79]. Furthermore, chondrocyte-like and osteocytes-like cells are reported to co-exist in chondroid bone, with no clearly distinguishable boundary. According to the examination results of BMP-induced heterotopic bone tissue, chondrocyte-like cells demonstrate the two-phase function of the chondrocytes and osteocytes. These data also suggest that this BMP induced cartilage-like tissue should be classified as chondroid bone rather than as normal cartilage. As mentioned above, these examination data suggest that the cells involved in chondroid bone express cartilage phenotypes temporarily, then become bone-forming cells which survive in chondroid bone until tissue is resorped and remodeled by true bone tissue, although there is no direct evidence of this [38]. Using immunohistochemical and in situ hybridization examinations, there are some data that transforming growth factor-β peptide may be involved in the differentiation of chondrocytes into bone-forming cells [23,68]. BMP-induced bone tissue is formed by the third ossification mode, "transchondroid bone formation".

Concerning the "chondroid bone or chondroid bone forming cells", they also appear in the neoplastic lesions of bone. In osteochondromas, cartilage-capped bone tissue projections are typical histopathological features. The surface of the masses is covered with a cartilage tissue showing positive immunohistochemical reaction for type II collagen, and the deep region is composed of spongy bone, showing a positive immunohistochemical reaction for type II collagen and osteocalcin. Between the cartilage and spongy bone, which is a metaphysic-like region, a chondroidal pattern appeared in the matrix of hypertrophy cartilage. In these regions, both type I and II collagens and osteosarcoma are immunohistochemically detected. Furthermore, type I collagen mRNA in neoplastic chondrocytes is detected. Therefore, the cells involved in "chondroid bone" appear in osteochondromas and then change directly into bone-forming cells that survive in the "chondroid bone" until the tissue is resorbed and remodeled into true bone tissue. These features suggest that bone formation in osteochondromas, at least in some regions, occurs through transchondroid bone formation.

Chapter 13

CONCLUSION

In the oral and craniofacial region, development is a complex and fascinating set of processes which require a sequential integration of numerous biological steps, and the differentiation in cells of neoplasms is also a complex and fascinating compound processes. In this region, there are many different types of neoplasms. It is known that proliferation, development and cytological-differentiation of the neoplastic cells reflect the normal physiological development of the outbreak mother cells and/or tissues. Therefore, some cell-differentiation, development and proliferation factors may also play some roles in the neoplastic cells, and therefore, their behavior is closely related in cytological differentiation and clinical behavior and/or grade.

REFERENCES

[1] Andela VB, Siddiqui F, Groman A and Rosier RN (2005) An immunohistochemical analysis to evaluate an inverse correlation between Runx2/Cbfa1 and NF kappa B in human osteosarcoma. *J Clin Pathol 58*: 328-330.

[2] Barsky SH, Siegal GP, Jannotta F and Liotta LA (1983) Loss of basement membrane components by invasive tumors but not by their benign counterparts. *Lab Invest 49*: 140-147.

[3] Bernfield M, Gotte M, Park PW, Reizes O, Fitzgerald ML, Lincecum J and Zako M (1999) Functions of cell surface heparin sulfate proteoglycans. *Ann Rev Biochem 68*: 729-777.

[4] Blackhall FH, Merry CL, Davies EJ and Javson GC (2001) Heparan sulfate proteoglycans and cancer. *Br J Cancer 85*: 1094-1098.

[5] Blaumueller CM, Qi H, Zagouras P, and Artavanis-Tsakonas S (1997) Intracellular cleavage of Notch leads to a heterodimeric receptor on the plasma membrane. *Cell 90*: 281-291.

[6] Butler WT and Ritchie H (1995) The nature and functional significance of dentin extracellular matrix proteins. *Int J Dev Biol 39*: 169-179.

[7] Dehan P, Waltregny D, Baschin A, Noel A, Castronovo V, Tryggvason K, Leval DJ and Foidart JM (1997) Loss of type IV collagen alpha 5 and alpha 6 chains in human invasive prostate carcinomas. *Am J Pathol 151*: 1097-1104.

[8] D'errico JA, Macneil RL, Takata T, Berry J, Strayorn C and Somerman MJ (1997) Expression of bone associated makers by tooth root lining cells, in situ and in vitro. *Bone 20*: 117-126.

[9] Fujii H, Nagatsuka H, Lee YJ, Shinnou T, Tamamura R, Xiao J, Naitou I, Sado G, Nakagawa T and Nagai N (2004) Differential expression of type IV

collagen alpha 1 to 6 chains in basement membrane of human tooth germ and odontogenic tumors. *J Hard Tissue Biol 13*: 103-109.

[10] Fujii H, Nagatsuka H, Lee YJ, Shinnou T, Tamamura R, Xiao J, Naitou I, Sado G, Nakagawa T and Nagai N (2004) Differential expression of type IV collagen alpha 1 to alpha 6 chains in basement membrane of malignant odontogenic tumors. *J Hard Tissue Biol 13*: 111-115.

[11] Gardner DG, Heikinheimo K, Shear M, Philipsen HP and Coleman H (2005) Ameloblastomas. In: Banes L, Eveson JW, Reichart P, Sidransky D ed. *World Health Organization Classification of Tumours Pathology and Genetics of Head and Neck Tumours.* IARC Press, Lyon, 296-300.

[12] Hamada T, Suda N and Takayuki K (1999) Immunohistochemical localization of fibroblast growth factor receptors in the rat mandibular condylar cartilage and tibial cartilage. *J Bone Miner Metab 17*: 274-282.

[13] Han PP, Nagatsuka H, Siar CH, Tsujigiwa H, Gunduz M, Tamamura R, Katase N, Nakajima M, Naomoto Y and Nagai N (2006) Immunodetection of heparin sulphate and heparanase molecules in benign and malignant odontogenic tumors. *Oral Med Pathol 11*: 49-54.

[14] Harada H, Mitsuyasu T and Toyono T (2002) Epithelial stem cells in teeth. *Odontol 90*: 1-6.

[15] Heikinheimo K, Jee KJ, Niini T, Aalto Y, Happonen RP, Leivo I and Knuutila S (2002) Gene expression profiling of ameloblastoma and human tooth germ by means of a cDNA microarray. *J Dent Res 81*: 525-30.

[16] Heikinheimo K, Morgan PR, Happonen RP, Stenman G and Virtanen I (1991) Distribution of extracellular matrix protein in odontogenic tumors and developing teeth. *Virchows Arch 61*: 101-109.

[17] Heikinheimo K and Salo T (1995) Expression of basement membrane type IV collagen and type IV collagenase (MMP-2 and MMP-9) in human fetal teeth. *J Dent Res 74*: 1226-1234.

[18] Hina M, Inoue M, Nagatsuka H, Takagi T and Nagai N (1993) Immunohistochemical demonstration of amelogenin, type IV collagen and fibronectin in human and rat tooth germs. *J Jpn Stomatol 42*: 659-664.

[19] Hudson BG, Reeders ST and Tryggvason K (1993) Type IV collagen: structure, gene organization, and role in human diseases. Molecular basis of Goodpasture and Alport syndromes and diffuse leiomyomatosis. *J Biol Chem 15*: 26033-26036.

[20] Ishida O, Inomata K and Nagai N (1992) Cytokeratin filaments in odontogenic epithelium of human fetus. *J Jpn Stomatol 41*: 19-29.

[21] Ishii M, Suda N, Toshimoto T, Shoichi S and Kuroda T (1998) Immunohistochemical findings type I and type II collagen in prenatal mouse

mandibular condylar cartilage compared with the tibial anlage. *Arch Oral Biol 43*: 545-550

[22] Kawakami T (2001) Immunohistochemistry of BMP-induced hetorotopic osteogensis. *J Hard Tissue Biol 10*: 73-76.

[23] Kawakami T, Hiraoka BY, Kawai T, Takei N, Hasegawa H and Eda S (1999) Expression of transforming growth factor-beta peptide and its mRNA in chondrocytes in the early phase of BMP-induced heterotopic 'transchondroid bone formation'. *Med Sci Res 27*: 419-421.

[24] Kawakami T, Kawai T, Kimura A, Hasegawa H, Tsujigiwa H, Gunduz M, Nagatsuka H and Nagai N (2001) Characteristics of bone morphogenetic protein-induced chondroid bone: histochemical, immunohistochemical and in situ hybridization examinations. *J Int Med Res 29*: 480-487.

[25] Kawakami T, Kawai T, Kimura A, Hasegawa H, Yoshikawa Y and Eda S (1998) Transchondroid bone formation displayed in BMP-induced hetrotopic osteogenesis. *J Hard Tissue Biol 7*: 21-26.

[26] Kawakami T, Kawai T, Takei N, Kise T, Eda S and Urist MR (1997) Evaluation of heterotopic bone formation induced by squalane and bone morphogenetic protein composite. *Clin Orthop 337*: 261-266.

[27] Kawakami T, Kimura A, Yamada M, Matsuura S, Horio T, Hasegawa H and Kanda H (2002) Localization of matrix proteins of hard tissue in osteosarcomas. *Eur J Med Res 7*: 335-339.

[28] Kawakami T, Shimizu M and Shimizu T (2005) Immunohistochemical characteristics of developing mandibular angle in fetal mice. *Eur J Med Res 10*: 547-548.

[29] Kawakami T. Shimizu T, Kimura A, Hasegawa H, Siar CH, Ng KH, Nagatsuka H, Nagai N and Kanda H (2005) Immunohistochemical examination of cytological differentiation in osteosarcomas. *Eur J Med Res. 10*: 475-479.

[30] Kawakami T, Siar CH, Ng KH, Shimizu T, Okafuji N, Kurihara S, Hasegawa H, Tsujigiwa H, Nagatsuka H and Nagai N (2004) Expression of Notch in a case of osteosarcoma of the maxilla. *Eur J Med Res 9*: 533-535.

[31] Kawakami T, Uji H, Antoh M, Hasegawa, Kise T and Eda S (1993) Squalane as a possible carrier of bone morphogenetic protein. *Boimateials 14*: 575-577.

[32] Kimura A, Kawakami T, Matsuura S, Hasegawa H, Kanda H, Tsujigiwa H, Nagatsuka N and Nagai N (2003) Gene expression of type I collagen in neoplastic chondrocytes. *Eur J Med Res 8*: 165-167.

[33] Kuboki T, Kanyama M, Nakanishi T, Akiyama K, Nawachi K, Yatani H, Yamashita K, Takano-Yamamoto T and Takigawa M (2003) Cbfa1/Runx2

gene expression in articular chondrocytes of the mice temporomandibular and knee joints in vivo. *Arch Oral Biol. 48*(7): 519-525

[34] Kumamoto H, Ohki K and Ooya K (2004) Expression of sonic hedgehog (SHH) signaling molecules in ameloblastomas. *J Oral Pathol Med 33*: 185-190.

[35] Mitsiadis TA, Fried K and Goridis C (1995) Reactivation of Delta-Notch signaling after injury: complementary expression pattern of ligand and receptor in dental pulp. *Exp Cell Res 246*: 312-318.

[36] Mitsiadis TA, Henrique D, Thesleff I and Lendahl U (1997) Mouse Serrate-1 (Jagged-1): expression in the developing tooth is regulated by epithelial-mesenchymal interactions and fibroblast growth factor-4. *Development 124*: 1473-1483.

[37] Mizoguchi I, Nakamura M, Takahashi I, Kagayama M and Mitani H (1990) An immunohistochemical study of localization of type I and type II collagens in mandibular condylar cartilage compared with tibial growth plate. *Histochemistry 93* (6): 593-599

[38] Nagai N, Nagatsuka H, Murata M, Inoue M, Akagi T, Qin C-L, Nakano K, Ishiwari Y, Konouchi H, Tsujigiwa H, Chigono Y and Takagi T (1995) Gene expression of bone matrix protein mRNA during BMP induced chondrogenesis and osteogenesis by in situ hybridization. *J Hard Tissue Biol 4*: 15-23.

[39] Nagai N, Nakano K, Sado Y, Naito I, Gunduz M, Tsujigiwa H, Nagatsuka H, Ninomiya Y and Siar CH (2001) Localization of type IV collagen a1 to a6 chains in basement membrane during mouse molar germ development. *Int J Dev Biol 45*: 827-831.

[40] Nagai N, Yamachika E, Nishijima K, Inoue M, Shin HI, Suh MS, Nagatsuka H (1994) Immunohistochemical demonstration of tenascin and fibronectin in odontogenic tumors and human fetal tooth germs. *Oral Oncol, Eur J Cancer 30B*: 191-195.

[41] Nagatsuka H, Han PP, Tsujigiwa H, Siar CH, Gunduz M, Sugahara T, Sasaki A, Nakajima M, Naomoto Y and Nagai N (2005) Heparanase gene and protein expression in ameloblastoma: possible role in local invasion of tumor cells. *Oral Oncol 41*: 542-548.

[42] Nagatsuka H, Shin H-I, Park H-K, Ishiwari Y, Kuroda K, Nosaka Y, Song H, Qin C-L, Zhang S-Q, Nakano K, Chigono Y, Tsujigiwa H, Takagi T and Nagai N (1995) Immunohistochemical study of cytokeratin patterns in follicular and plexiform ameloblastoma. *J Hard Tissue Biol 4*: 50-56.

[43] Nagatsuka H, Siar CH, Kitamura Y, Tsujigiwa H, Lee Y-J, Gunduz M, Tamamura R, Komori T, Lefevre M and Nagai N (2004) Gene expression of matrix proteins in Cbfa1-knockout mice. *J Hard Tissue Biol 13*: 37-46.

[44] Nagatsuka H, Siar CH, Nakano K, Tsujigiwa H, Gunduz M, Choufuku H, Lee Y-J, Naito I, Sado Y and Nagai N (2002) Differential expression of collagen IV α1 to α6 chains in basement membranes of benign and malignant odontogenic tumors. *Virch Arch 441*: 392-399.

[45] Nakano K, Siar CH, Nagai N, Naito I, Sado Y, Nagatsuka H, Hor C, Kuroda K, Tsujigiwa H and Gunduz M (2002) Distribution of basement membrane type IV collagen a chains in ameloblastoma: an immunofluoressence study. *J Oral Pathol Med 31*: 494-499.

[46] Nakano S, Iyama K, Ogawa M, Yoshioka H, Sado Y, Oohashi T and Ninomiya Y (1999) Differential tissular expression and localization of type IV collagen alpha1 (IV), alpha2 (IV), alpha5 (IV), and alpha6 (IV) chains and their mRNA in normal breast and in benign and malignant breast tumors. *Lab Invest 79*: 281-292.

[47] Ng KH and Siar CH (1990) Peripheral ameloblatoma with clear cell differentiation. *Oral Surg 70*: 210-213.

[48] Ninomiya Y, Kagawa M, Iyama K, Naito I, Kishiro Y, Seyer JM, Sugimoto M, Oohashi T and Sado Y (1995) Differential expression of two basement membrane collagen genes, COL4A6 and COL4A5, demonstrated by immunofluorescence staining using peptide-specific monoclonal antibodies. *J Cell Biol 130*: 1219-1229.

[49] Nobta M, Tsukazaki T, Shibata Y, Xin C, Moriishi T, Sakano S, Shindo H and Yamaguchi A (2005) Critical regulation of bone morphogenetic protein-induced osteoblastic differentiation by Delta1/Jagged1-activated Notch1 signaling. *J Biol Chem 280*: 15842-15848

[50] Parish CR, Freeman C and Hulett MD (2001) Heparanase: a key enzyme involved in cell invasion. *Biocem Biophys Acta 1471*: M99-108.

[51] Philipsen HP, Ormiston IW and Reichart PA (1992) The desmo and osteoplastic ameloblastoma. Histologic valiant or clinicopathoilogic entity? Case reports. *Int J Oral Maxillofac Surg 21*: 352-357.

[52] Rabie ABM and Hagg U (2002) Factors regulating mandibular condylar growth. *Am J Orthod Dentofacial Orthop 122*: 401-409.

[53] Reichart PA, Philipsen HP and Sonner S (1995) Ameloblastoma: biological profile of 3677 cases. *Eur J Cancer B Oral Oncol 31B*: 86-99.

[54] Sado Y, Kagawa M, Kishiro Y, Sugihara K, Naito I, Seyer JM, Sugimoto M and Oohashi T (1995) Establishment by the rat lymphnode method of

epitope-defiened monoclonal antibodies recognizing the six different a chains of human type IV collagen. *Histochem Cell Biol 104*: 267-275.

[55] Sado Y, Kagawa M, Naito I, Ueki Y, Seki T, Momota R, Oohashi T and Ninomiya Y (1998) Organization and expression of basement membrane collagen IV genes and their roles in human disorders. *J Biochem 123*: 767-776.

[56] Schnabel M, Fichtel I, Gotzen L and Schlegel J (2002) Differential expression of Notch genes in human osteoblastic cells. *Int J Molecul Med 9*: 229-232

[57] Shibata S, Fukuda K, Suzuki S, Ogawa T and Yamashita Y. (2002) In situ hybridization and immunohistochemistry of bone sialoprotein and secreted phosphoprotein 1 (osteopontin) compared with limb bud cartilage. *J Anat 200*: 309-320

[58] Shibata S, Suda N, Suzuki S, Fukuoka H and Yamasita Y (2006) An in situ hybridization study of Runx2, Osterix, and Sox9 at the onset of condylar cartilage formation in fetal mouse mandible. *J Anat 208*: 169-177

[59] Shibata S, Suda N, Yoda S, Fukuoka H, Ohyama K, Yamashita Y, Komori T (2004) Runx2-deficient mice lack mandibular condylar cartilage. *Anat Embryol 208*(4): 273-80

[60] Shibata S, Suzuki S, Tengan T, Ishii M and Kuroda T (1996) A histological study of the developing condylar cartilage of the fetal mouse mandible using coronal sections. *Arch Oral Biol 41*(1): 47-54

[61] Shimizu M (2005) Histological and immunohistochemical observations of developing mandibular angle in mice. *J Hard Tissue Biol 14*: 346-350.

[62] Shimizu T (2005) Histological characteristics of mandible in fetal mice. *J Hard Tissue Biol 14*: 67-68.

[63] Shimizu T (2006) Participation of Runx2 in mandibular condylar cartilage development. *Eur J Med Res 11*: 455-461.

[64] Shimizu T, Tsujigiwa H, Nagatsuka H, Okafuji N, Kurihara S, Nagai N and Kawakami T (2005) Expression of Notch1 and Math1 in mandibular condyle cartilage in neonatal mice. *Angl Orthodont 75*: 993-995.

[65] Siar CH and Ng KH (1991) Calcifying and keratinizing ameloblastoma of the maxilla. *J Laryngol Otol 105*: 971-972.

[66] Siar CH, Ng KH, Ariff Z, Muraki E, Shimizu T, Tsujigiwa H, Nagatsuka H, Nagai N and Kawakami T (2006) A case report of ameloblastoma of the mandible with examination of Notch signaling. *Oral Med Pathol 11*: 35-39.

[67] Slater LJ (1999) Odontogenic sarcomas and carcinosarcomas. *Semin Diagn Pathol 16*: 325-332.

References

[68] Takei N, Kawakami T, Kawai T, Yoshikawa Y and Eda (1997) Expression of TGF-β in the course of BMP-induced heterotopic osteogenesis. *J Hard Tissue Biol 6*: 64-69.

[69] Tamamura R, Nagatsuka H, Lee Y-J, Xiao J, Naito I, Sado Y, Kawakami T and Nagai N (2004) Distribution of collagen type IV α1-6 chains and MMP in carcinogenesis in oral squamous cell carcinoma. *J Hard Tissue Biol 13*: 117-123.

[70] Tamamura R, Nagatsuka H, Lee Y-J, Xiao J, Naito I, Sado Y, Kawakami T and Nagai N (2004) Distribution of collagen type IV α1-6 chains and MMP in oral squamous cell carcinoma. *J Hard Tissue Biol 13*: 1125-130.

[71] Tanaka K, Iyama K, Kitaoka M, Ninomiya Y, Oohashi T, Sado Y and Ono T (1997) Differential expression of alpha 1 (IV), alpha 2 (IV), alpha 5 (IV) and alpha 6 (IV) collagen chains in the basement membrane of basal cell carcinoma. *Histochem J 29*: 563-570.

[72] Tezuka K, Yasuda M, Watanabe N, Morimura N, Kuroda K, Miyatani S and Hozumi N (2002) Stimulation of osteoblastic cell differentiation by Notch. *J Bone Miner Res 17*: 231-239.

[73] Thesleff I, Barrach HJ, Foidart JM, Vaheri A, Pratt RM and Martin GR (1981) Changes in the distribution of type IV collagen, laminin, proteoglycan, and fibronectin during mouse tooth development. *Dev Biol 81*: 182-192.

[74] Thesleff I and Ekblom P (1984) Distribution of keratin and laminin in ameloblastoma. Comparison with developing tooth and epidermoid carcinoma. *J Oral Pathol 13*: 85-96.

[75] Thomas DM, Johnson SA, Sims NA, Trivett MK, Slavin JL, Rubin BP, Waring P, MacArthur GA, Walkley CR, Holloway AJ, Diyagama D, Grim JE, Clurman BE, Bowtell DD, Lee JS, Gutierrez GM, Piscopo DM, Carty SA and Hinds PW (2004) Terminal osteoblast differentiation, mediated by Runx2 and p27LIP1, is disrupted in osteosarcoma. *J Cell Biol 167*: 925-934

[76] Tomich CE (1999) Benign mixed odontogenic tumors. *Semin Diagn Pathol 16*: 308-316.

[77] Tsujigiwa H, Nagatsuka H, Han PP, Gunduz M, Siar CH, Oida S and Nagai N (2005) Analysis of amelogenin gene (AMGX, AMGY) expression in ameloblastoma. *Oral Oncol 41*: 843-850.

[78] Watanabe N, Tezuka Y, Matsuno K, Miyatani S, Morimura N, Yasuda M, Fujimaki R, Kuroda K, Hiraki Y, Hozumi N, Tezuka K (2003) Suppression of differentiation and proliferation of early chondrogenic cells by Notch. *J Bone Miner Metab 21*: 344-352

[79] Yasui N, Sato M, Ochi T, Kimura T, Kawahara H, Kitamura M and Nomura S (1997) Tree modes of ossification during destruction osteogenesis in the rat. *J Bone Joint Surg 79-B*: 824-830.

[80] Yurchenco PD and O'Rear JJ (1994) Basal lamina assembly. *Curr Opin Cell Biol 6*: 674-681.

[81] Zagouras P, Stifani S, Blaumueller CM, Carcangiu ML and Artavanis-Tsakonas S (1995) Alterations in Notch signaling in neoplastic lesions of the human cervix. *Proc Natl Acad Sci USA 92*: 6414-6418.

[82] Zcharia E, Metzger S, Chajek-Shaul T, Aingorn H, Elkin M, Friedmann Y, Weinstein T, Li JP, Lindahl U and Vlodavsky I (2004) Ttansgenic expression of mammalian heparanase uncovers physiological functions of heparansulfate in tissue morphogenesis, vascularization, and feeding behavior. *FASEB J 18*: 252-263.

INDEX

A

adhesion, 30
adult, 32
alpha, 43, 44, 49
amorphous, 28
animals, 30
articular cartilage, 12, 13
articulation, 11
attention, 23, 39

B

basal cell carcinoma, 49
basement membrane, 43, 44, 46, 47, 48, 49
behavior, 41, 50
bell, 5, 7, 8
benign, 3, 21, 24, 25, 26, 28, 29, 30, 31, 34, 43, 44, 47
binding, 8, 30
biochemical, 12
biological, vii, 9, 22, 24, 30, 33, 41, 47
birth, 12, 14, 15
BMPs, 37
bone, 5, 8, 9, 11, 12, 14, 15, 19, 33, 34, 35, 37, 38, 39, 40, 43, 45, 46, 47, 48
bone marrow, 39
breast, 47
budding, 30

C

cancer, iv, 29, 43, 46, 47
cancer cells, 29
carcinogenesis, 49
carcinoma, 25, 29, 31, 49
carcinomas, 23, 25, 28, 30, 31, 43
carrier, 45
cartilage, 9, 11, 12, 13, 14, 15, 19, 37, 38, 39, 40, 44, 45, 46, 48
cartilaginous, 11, 14
cDNA, 24, 44
cell, vii, 1, 3, 5, 6, 8, 9, 11, 12, 13, 15, 17, 18, 19, 22, 23, 24, 25, 26, 27, 28, 29, 30, 31, 32, 33, 35, 41, 43, 47, 49
cell adhesion, 30
cell cycle, 35
cell differentiation, 3, 19, 28, 35, 47, 49
cell division, 17
cell fate, vii, 17, 18, 24, 26
cell growth, 29
cell invasion, 47
cell surface, 30, 43
cellular atypia, 25, 26
cellular regulation, 9
cementum, 5
cervix, 50
chondrocyte, 14, 39

chondrocytes, 12, 13, 14, 15, 37, 39, 40, 45, 46
chondrogenic, 49
chromosomes, 27
classification, 21
classified, 11, 21, 40
cleavage, 43
clinical, 11, 30, 41
coagulation, 12, 13
collagen, 7, 12, 13, 14, 27, 28, 38, 39, 40, 43, 44, 45, 46, 47, 48, 49
complementary, 46
components, 1, 3, 7, 8, 9, 12, 27, 28, 31, 43
composite, 45
condensation, 11, 13, 15
confinement, 29
Congress, iv
connective tissue, 8, 26, 27, 31
continuity, 29
control, 1, 6, 15, 33, 35, 36
controlled, 3, 33
correlation, 27, 43
coupling, 35
covering, 5
craniofacial, iv, vii, 9, 11, 19, 26, 41
cyst, 23, 26
cytokeratins, 32
cytokines, 30
cytoplasm, 13, 14, 15, 23, 26, 32, 33, 35, 39

D

deciduous, 5
Delta, 34, 35, 46
dentin, 5, 7, 43
deposits, 28
destruction, 50
differentiated cells, 3
differentiation, iv, vii, 1, 3, 5, 6, 8, 12, 14, 15, 17, 19, 23, 24, 26, 27, 28, 31, 32, 33, 35, 36, 39, 41, 45, 47, 49
discontinuity, 27
displacement, 15
distal, 11, 13, 15

distribution, 7, 8, 13, 15, 17, 23, 24, 27, 31, 32, 49
division, 17
doctors, vii

E

ectoderm, vii, 5, 10
electronic, iv
electrostatic, iv
embryo, 9
embryonic, 9, 11, 13, 14, 15
endoderm, vii, 10
enzyme, 47
epigenetic, 27
epithelial cell, 17, 28, 30, 31, 32
epithelial cells, 17, 30, 32
epithelium, 5, 6, 7, 8, 17, 23, 24, 32, 44
epitope, 48
evidence, 15, 37, 40
examinations, 13, 14, 26, 40, 45
expert, iv
extracellular, 12, 14, 15, 30, 31, 43, 44
extracellular matrix, 12, 14, 30, 31, 43, 44

F

failure, 3
family, 35
feeding, 50
fetal, 7, 24, 32, 44, 45, 46, 48
fetus, 44
fibrillar, 31, 39
fibroblast, 13, 44, 46
fibroblast growth factor, 13, 44, 46
fibroma, 25, 31
fibronectin, 7, 8, 31, 44, 46, 49
fibrous tissue, 32
fish, 9
follicle, 8, 31
follicular, 23, 25, 26, 31, 32, 46
Friedmann, 50

G

gene, 6, 12, 15, 24, 25, 26, 27, 38, 39, 44, 46, 49
gene expression, 6, 12, 15, 24, 25, 26, 38, 39, 46
generation, 13
genes, 12, 23, 25, 27, 35, 47, 48
glycoprotein, 8
grades, 31
growth, 1, 3, 8, 12, 13, 19, 24, 27, 29, 30, 31, 40, 44, 45, 46, 47
growth factor, 13, 30, 31, 40, 44, 45, 46
growth factors, 30, 31

H

hair follicle, 31
histochemical, 45
histological, 11, 30, 32, 48
host, 29
human, vii, 8, 9, 24, 30, 31, 35, 43, 44, 46, 48, 50
hybridization, 14, 40, 45, 46, 48
hypertrophy, 12, 13, 14, 15, 40

I

IARC, 44
immature cell, 17
immunofluorescence, 47
immunohistochemical, vii, 13, 14, 29, 31, 32, 34, 35, 40, 43, 45, 46, 48
immunohistochemistry, 23, 26, 48
immunoreactivity, 32
in situ, 14, 40, 43, 45, 46, 48
in situ hybridization, 14, 40, 45, 46, 48
in vitro, 43
in vivo, 46
incisor, 6
initiation, 5
injury, iv, 46
integration, vii, 9, 41
intensity, 23, 28
interaction, 17, 33
interactions, 5, 6, 8, 9, 17, 24, 30, 32, 46
interface, 8, 28
invasive, 30, 43
Investigations, 13
isoforms, 7
IV collagenase, 44

J

joints, 46

K

kappa, 43
kappa B, 43
keratin, 24, 49
knockout, 5, 47

L

labeling, 24
lamellar, 39
lamina, 7, 8, 32, 50
laminin, 7, 49
lesions, 3, 40, 50
ligament, 10
ligand, 17, 35, 46
linear, 27, 28, 32
literature, 13, 22, 24, 35
localization, 8, 28, 29, 31, 34, 38, 44, 46, 47
location, 30
lumen, 30

M

macromolecules, 30
magnetic, iv
malignancy, 3, 30
malignant, 3, 8, 24, 25, 26, 28, 30, 31, 34, 44, 47
malignant tumors, 25, 34
mandible, 14, 48

mandibular, 9, 11, 12, 13, 14, 15, 44, 45, 46, 47, 48
marrow, 39
matrix, 5, 8, 12, 14, 27, 30, 31, 33, 39, 40, 43, 44, 45, 46, 47
matrix protein, 5, 8, 14, 27, 39, 43, 44, 45, 46, 47
maturation, 6, 14
maxilla, 35, 45, 48
maxillary, 9
mechanical, iv
membranes, 47
mesenchymal, 5, 6, 9, 11, 13, 14, 15, 17, 24, 30, 32, 33, 37, 39, 46
mesenchyme, 5, 6, 8, 10, 17
mesoderm, vii, 5, 10
mice, 5, 7, 37, 39, 45, 46, 47, 48
microarray, 24, 44
mineralized, 28
mitosis, 25, 26, 33
MMP, 44, 49
MMP-2, 44
MMP-9, 44
modulation, 31
molecules, 7, 28, 29, 30, 44, 46
monoclonal, 47, 48
monoclonal antibodies, 47, 48
morphogenesis, vii, 5, 6, 13, 17, 24, 26, 31, 33, 34, 50
morphological, 1, 5, 8, 12
morphology, 8, 14
mother cell, vii, 3, 26, 41
mouse, 7, 11, 12, 17, 44, 46, 48, 49
mRNA, 12, 15, 24, 25, 27, 38, 39, 40, 45, 46, 47
mucosa, 32
muscle, 14
muscle tissue, 14

N

necrosis, 24
neonatal, 48
neoplasm, vii, 1, 3, 21, 28, 32

neoplasms, v, vii, 1, 3, 8, 19, 21, 22, 24, 26, 27, 28, 29, 30, 31, 33, 41
neoplastic, iii, iv, vii, 1, 3, 8, 19, 24, 25, 26, 27, 28, 29, 30, 33, 34, 35, 40, 41, 45, 50
neoplastic cells, iv, vii, 24, 25, 26, 30, 33, 35, 41
neural crest, vii, 5, 9
New York, iii, iv
normal, vii, 3, 12, 13, 14, 27, 37, 40, 41, 47
normal development, 14
NSE, 32
nuclear, 29
nuclei, 23, 29, 30, 33
nucleus, 14, 15, 23

O

observations, 48
odontogenic, vii, 6, 8, 17, 19, 21, 23, 24, 28, 29, 30, 31, 32, 44, 46, 47, 49
odontogenic tumors, 28, 30, 44, 46, 47, 49
oncogenesis, 35
oral, iv, vii, 5, 7, 9, 11, 19, 26, 27, 30, 32, 41, 49
oral cavity, 5, 7
oral squamous cell carcinoma, 30, 49
organ, vii, 8, 26, 32
organization, 3, 12, 44
ossification, 12, 13, 14, 15, 37, 38, 39, 50
osteoblastic cells, 48
osteoblasts, 6, 33, 39
osteocalcin, 39, 40
osteocytes, 39, 40
osteogenic, 8, 31
osteopontin, 6, 12, 13, 14, 33, 48
osteoporosis, 35
osteosarcoma, vii, 33, 34, 35, 40, 43, 45, 49

P

parenchyma, 24
parenchymal, 1, 3, 23, 26
parenchymal cell, 1, 3, 23
pathology, 35

pathways, 6, 31
peptide, 7, 12, 14, 15, 23, 25, 33, 35, 37, 40, 45, 47
peptides, 24
periodontal, 32
periodontium, 8
Peripheral, 47
pharynx, 10
phenotypes, 37, 40
phosphoprotein, 48
physiological, vii, 1, 3, 8, 12, 13, 19, 26, 27, 30, 35, 37, 41, 50
physiological regulation, 3
physiology, iv
physiopathology, iv
plasma, 43
plasma membrane, 43
platelet, 13
play, vii, 8, 13, 24, 41
preparation, iv
production, 33
progenitor cells, 33
progeny, 17
proliferation, vii, 9, 11, 13, 15, 19, 23, 24, 28, 33, 39, 41, 49
promote, 31
property, iv
prostate, 43
prostate carcinoma, 43
protein, 6, 8, 14, 27, 30, 39, 44, 45, 46, 47
proteins, 5, 30, 43, 45, 47
proteoglycans, 30, 43
pulp, 7, 46

R

random, 27
range, 32
rat, 44, 47, 50
reactivity, 28, 29, 31, 32
receptors, 44
regulation, vii, 3, 6, 9, 13, 15, 17, 24, 26, 35, 47
regulators, 33, 34
relationship, vii, 26, 35

remodeling, 29
research, 14
researchers, 11
reticulum, 17, 23, 28, 32
RNA, 15

S

sarcomas, 48
scientific, 9
series, 24
services, iv
sex, 27
sex chromosome, 27
shape, 33
signaling, vii, 5, 6, 13, 14, 17, 18, 23, 24, 26, 31, 35, 46, 47, 48, 50
signaling pathway, 5, 6, 31
signaling pathways, 5, 31
signals, 15, 24, 25, 38
sites, 37, 39
specialization, 3
specialized cells, 3
speculation, 23
spindle, 32, 33, 39
squamous cell, 30, 32, 49
squamous cell carcinoma, 30, 49
stages, 6
stellate cells, 30
stem cells, 3, 17, 33, 44
strength, 24, 35, 36
stroma, 23, 27, 31
stromal, 8, 21, 24, 28, 29, 30, 31
stromal cells, 30
sugar, 30, 31
sulfate, 7, 43
sulphate, 29, 44
suppressor, 35
surgical, 27

T

teeth, 5, 6, 44
tenascin, 8, 31, 46

TGF, 24, 37, 38, 49
time, 15
tissue, vii, 1, 3, 5, 8, 9, 14, 17, 21, 25, 26, 27, 28, 29, 30, 31, 33, 34, 35, 36, 37, 38, 39, 40, 45, 50
trabecular bone, 12, 39
trans, 32
transcription, 5, 12, 14, 24
transcription factor, 5, 12, 14, 24
transformation, 8, 30, 31
transforming growth factor, 40, 45
transforming growth factor-β, 40
transmembrane, 17
tumor, 24, 27, 28, 30, 35, 46
tumor cells, 28, 46
tumor growth, 27
tumorigenesis, 27
tumo(u)rs, iv, 21, 25, 28, 30, 34, 43, 44, 46, 47, 49

U

ubiquitous, 30
undifferentiated, 37, 39
undifferentiated cells, 39
undifferentiated mesenchymal cells, 37

V

values, 9
variable, 24, 33
variation, 24
vascular, 7
vascularization, 50
visible, 6, 12, 25, 34, 38, 39

W

withdrawal, 35
World Health Organization (WHO), 21, 44

Soviet and Post-Soviet Politics and Society (SPPS)
ISSN 1614-3515

Founded in 2004 and refereed since 2007, SPPS makes available affordable English-, German-, and Russian-language studies on the history of the countries of the former Soviet bloc from the late Tsarist period to today. It publishes between 5 and 20 volumes per year and focuses on issues in transitions to and from democracy such as economic crisis, identity formation, civil society development, and constitutional reform in CEE and the NIS. SPPS also aims to highlight so far understudied themes in East European studies such as right-wing radicalism, religious life, higher education, or human rights protection. The authors and titles of all previously published volumes are listed at the end of this book. For a full description of the series and reviews of its books, see www.ibidem-verlag.de/red/spps.

Editorial correspondence & manuscripts should be sent to: Dr. Andreas Umland, Institute for Euro-Atlantic Cooperation, vul. Volodymyrska 42, off. 21, UA-01030 Kyiv, Ukraine

Business correspondence & review copy requests should be sent to: *ibidem* Press, Leuschnerstr. 40, 30457 Hannover, Germany; tel.: +49 511 2622200; fax: +49 511 2622201; spps@ibidem.eu.

Authors, reviewers, referees, and editors for (as well as all other persons sympathetic to) SPPS are invited to join its networks at
www.facebook.com/group.php?gid=52638198614
www.linkedin.com/groups?about=&gid=103012
www.xing.com/net/spps-ibidem-verlag/

Recent Volumes

167 *Oksana Kim*
The Effects and Implications of Kazakhstan's Adoption of International Financial Reporting Standards
A Resource Dependence Perspective
With a foreword by Svetlana Vlady
ISBN 978-3-8382-0987-6

168 *Anna Sanina*
Patriotic Education in Contemporary Russia
Sociological Studies in the Making of the Post-Soviet Citizen
With a foreword by Anna Oldfield
ISBN 978-3-8382-0993-7

169 *Rudolf Wolters*
Spezialist in Sibirien
Faksimile der 1933 erschienenen ersten Ausgabe
Mit einem Vorwort von Dmitrij Chmelnizki
ISBN 978-3-8382-0515-1

170 *Michal Vít, Magdalena M. Baran (eds.)*
Transregional versus National Perspectives on Contemporary Central European History
Studies on the Building of Nation-States and Their Cooperation in the 20th and 21st Century
With a foreword by Petr Vágner
ISBN 978-3-8382-1015-5

171 *Philip Gamaghelyan*
Conflict Resolution Beyond the International Relations Paradigm
Evolving Designs as a Transformative Practice in Nagorno-Karabakh and Syria
With a foreword by Susan Allen
ISBN 978-3-8382-1057-5

172 *Maria Shagina*
Joining a Prestigious Club
Cooperation with Europarties and Its Impact on Party Development in Georgia, Moldova, and Ukraine 2004–2015
With a foreword by Kataryna Wolczuk
ISBN 978-3-8382-1084-1

173 *Alexandra Cotofana, James M. Nyce (eds.)*
Religion and Magic
in Socialist and Post-Socialist Contexts II
Baltic, Eastern European, and Post-USSR Case Studies
ISBN 978-3-8382-0990-6

# Soviet and Post-Soviet Politics and Society (SPPS)
ISSN 1614-3515

General Editor: Andreas Umland,
Institute for Euro-Atlantic Cooperation, Kyiv, umland@stanfordalumni.org

Commissioning Editor: Max Jakob Horstmann,
London, mjh@ibidem.eu

EDITORIAL COMMITTEE*

DOMESTIC & COMPARATIVE POLITICS
Prof. **Ellen Bos**, *Andrássy University of Budapest*
Dr. **Ingmar Bredies**, *FH Bund, Brühl*
Dr. **Andrey Kazantsev**, *MGIMO (U) MID RF, Moscow*
Prof. **Heiko Pleines**, *University of Bremen*
Prof. **Richard Sakwa**, *University of Kent at Canterbury*
Dr. **Sarah Whitmore**, *Oxford Brookes University*
Dr. **Harald Wydra**, *University of Cambridge*

SOCIETY, CLASS & ETHNICITY
Col. **David Glantz**, *"Journal of Slavic Military Studies"*
Dr. **Marlène Laruelle**, *George Washington University*
Dr. **Stephen Shulman**, *Southern Illinois University*
Prof. **Stefan Troebst**, *University of Leipzig*

POLITICAL ECONOMY & PUBLIC POLICY
Prof. em. **Marshall Goldman**, *Wellesley College, Mass.*
Dr. **Andreas Goldthau**, *Central European University*
Dr. **Robert Kravchuk**, *University of North Carolina*
Dr. **David Lane**, *University of Cambridge*
Dr. **Carol Leonard**, *Higher School of Economics, Moscow*
Dr. **Maria Popova**, *McGill University, Montreal*

FOREIGN POLICY & INTERNATIONAL AFFAIRS
Dr. **Peter Duncan**, *University College London*
Prof. **Andreas Heinemann-Grüder**, *University of Bonn*
Dr. **Taras Kuzio**, *Johns Hopkins University*
Prof. **Gerhard Mangott**, *University of Innsbruck*
Dr. **Diana Schmidt-Pfister**, *University of Konstanz*
Dr. **Lisbeth Tarlow**, *Harvard University, Cambridge*
Dr. **Christian Wipperfürth**, *N-Ost Network, Berlin*
Dr. **William Zimmerman**, *University of Michigan*

HISTORY, CULTURE & THOUGHT
Dr. **Catherine Andreyev**, *University of Oxford*
Prof. **Mark Bassin**, *Södertörn University*
Prof. **Karsten Brüggemann**, *Tallinn University*
Dr. **Alexander Etkind**, *University of Cambridge*
Dr. **Gasan Gusejnov**, *Moscow State University*
Prof. em. **Walter Laqueur**, *Georgetown University*
Prof. **Leonid Luks**, *Catholic University of Eichstaett*
Dr. **Olga Malinova**, *Russian Academy of Sciences*
Prof. **Andrei Rogatchevski**, *University of Tromsø*
Dr. **Mark Tauger**, *West Virginia University*

ADVISORY BOARD*

Prof. **Dominique Arel**, *University of Ottawa*
Prof. **Jörg Baberowski**, *Humboldt University of Berlin*
Prof. **Margarita Balmaceda**, *Seton Hall University*
Dr. **John Barber**, *University of Cambridge*
Prof. **Timm Beichelt**, *European University Viadrina*
Dr. **Katrin Boeckh**, *University of Munich*
Prof. em. **Archie Brown**, *University of Oxford*
Dr. **Vyacheslav Bryukhovetsky**, *Kyiv-Mohyla Academy*
Prof. **Timothy Colton**, *Harvard University, Cambridge*
Prof. **Paul D'Anieri**, *University of Florida*
Dr. **Heike Dörrenbächer**, *Friedrich Naumann Foundation*
Dr. **John Dunlop**, *Hoover Institution, Stanford, California*
Dr. **Sabine Fischer**, *SWP, Berlin*
Dr. **Geir Flikke**, *NUPI, Oslo*
Prof. **David Galbreath**, *University of Aberdeen*
Prof. **Alexander Galkin**, *Russian Academy of Sciences*
Prof. **Frank Golczewski**, *University of Hamburg*
Dr. **Nikolas Gvosdev**, *Naval War College, Newport, RI*
Prof. **Mark von Hagen**, *Arizona State University*
Dr. **Guido Hausmann**, *University of Munich*
Prof. **Dale Herspring**, *Kansas State University*
Dr. **Stefani Hoffman**, *Hebrew University of Jerusalem*
Prof. **Mikhail Ilyin**, *MGIMO (U) MID RF, Moscow*
Prof. **Vladimir Kantor**, *Higher School of Economics*
Dr. **Ivan Katchanovski**, *University of Ottawa*
Prof. em. **Andrzej Korbonski**, *University of California*
Dr. **Iris Kempe**, *"Caucasus Analytical Digest"*
Prof. **Herbert Küpper**, *Institut für Ostrecht Regensburg*
Prof. **Rainer Lindner**, *CEEER, Berlin*
Dr. **Vladimir Malakhov**, *Russian Academy of Sciences*

Dr. **Luke March**, *University of Edinburgh*
Prof. **Michael McFaul**, *Stanford University, Palo Alto*
Prof. **Birgit Menzel**, *University of Mainz-Germersheim*
Prof. **Valery Mikhailenko**, *The Urals State University*
Prof. **Emil Pain**, *Higher School of Economics, Moscow*
Dr. **Oleg Podvintsev**, *Russian Academy of Sciences*
Prof. **Olga Popova**, *St. Petersburg State University*
Dr. **Alex Pravda**, *University of Oxford*
Dr. **Erik van Ree**, *University of Amsterdam*
Dr. **Joachim Rogall**, *Robert Bosch Foundation Stuttgart*
Prof. **Peter Rutland**, *Wesleyan University, Middletown*
Prof. **Marat Salikov**, *The Urals State Law Academy*
Dr. **Gwendolyn Sasse**, *University of Oxford*
Prof. **Jutta Scherrer**, *EHESS, Paris*
Prof. **Robert Service**, *University of Oxford*
Mr. **James Sherr**, *RIIA Chatham House London*
Dr. **Oxana Shevel**, *Tufts University, Medford*
Prof. **Eberhard Schneider**, *University of Siegen*
Prof. **Olexander Shnyrkov**, *Shevchenko University, Kyiv*
Prof. **Hans-Henning Schröder**, *SWP, Berlin*
Prof. **Yuri Shapoval**, *Ukrainian Academy of Sciences*
Prof. **Viktor Shnirelman**, *Russian Academy of Sciences*
Dr. **Lisa Sundstrom**, *University of British Columbia*
Dr. **Philip Walters**, *"Religion, State and Society", Oxford*
Prof. **Zenon Wasyliw**, *Ithaca College, New York State*
Dr. **Lucan Way**, *University of Toronto*
Dr. **Markus Wehner**, *"Frankfurter Allgemeine Zeitung"*
Dr. **Andrew Wilson**, *University College London*
Prof. **Jan Zielonka**, *University of Oxford*
Prof. **Andrei Zorin**, *University of Oxford*

* While the Editorial Committee and Advisory Board support the General Editor in the choice and improvement of manuscripts for publication, responsibility for remaining errors and misinterpretations in the series' volumes lies with the books' authors.